NEUROSCIENCE
INTELLIGENCE
UNIT

Transplantation of Neural Tissue into the Spinal Cord
Second Edition

Antal Nógrádi, M.D., Ph.D.

Associate Professor
Department of Ophthalmology
University of Szeged
Albert Szent-Györgyi Medical Center
Szeged, Hungary

LANDES BIOSCIENCE / EUREKAH.COM
GEORGETOWN, TEXAS
U.S.A.

SPRINGER SCIENCE+BUSINESS MEDIA
NEW YORK, NEW YORK
U.S.A.

TRANSPLANTATION OF NEURAL TISSUE INTO THE SPINAL CORD
SECOND EDITION
Neuroscience Intelligence Unit

Landes Bioscience / Eurekah.com
Springer Science+Business Media, Inc.

ISBN: 0-387-26355-1   Printed on acid-free paper.

Springer Science+Business Media, Inc., 233 Spring Street, New York, New York 10013, U.S.A.
http://www.springeronline.com

Please address all inquiries to the Publishers:
Landes Bioscience / Eurekah.com, 810 South Church Street, Georgetown, Texas 78626, U.S.A.
Phone: 512/ 863 7762; FAX: 512/ 863 0081
http://www.eurekah.com
http://www.landesbioscience.com

Printed in the United States of America.

9 8 7 6 5 4 3 2 1

Library of Congress Cataloging-in-Publication Data

Nógrádi, Antal.
  Transplantation of neural tissue into the spinal cord / Antal Nógrádi.-- 2nd ed.
     p. ; cm. -- (Neuroscience intelligence unit)
  Rev. ed. of: Transplantation of neural tissue / Gerta Vrbová ... [et al.]. 1994.
  ISBN 0-387-26355-1
  1. Fetal nerve tissue--Transplantation. 2. Spinal cord--Regeneration. I. Transplantation of neural tissue into the spinal cord. II. Title. III. Series.
  [DNLM: 1. Nerve Tissue--transplantation. 2. Spinal Cord Injuries--surgery. 3. Nerve Regeneration. WL 400 N777t 2006]
  RD124.T732 2006
  617.4'820592--dc22
                                    2005016427

We dedicate this book to our families

CONTENTS

EDITOR

Antal Nógrádi
Associate Professor
Department of Ophthalmology
University of Szeged
Albert Szent-Györgyi Medical Center
Szeged, Hungary
Chapters 1, 4, 6, 7

CONTRIBUTORS

Gavin Clowry
Lecturer
School of Clinical Medical Sciences
University of Newcastle upon Tyne
Newcastle, United Kingdom
Chapter 2

Urszula Slawinska
Associate Professor
Department of Neurophysiology
Nencki Institute, Poland
Warsaw, Poland
Chapters 2, 3

Gerta Vrbová
Professor of Developmental Neuroscience
Department of Anatomy and Developmental Biology
University College London
London, United Kingdom
Chapters 1, 5

PREFACE

The book gives an account of results obtained from experiments where grafts of neuronal, glial and other tissues as well as artificial materials were placed into the spinal cord. It attempts to evaluate the contributions made by these studies to our understanding of basic neurobiological questions. These include factors that regulate neuronal growth during development as well as regeneration following injury to the nervous system. The model of neural transplantation is also useful for the study of cell-to-cell interactions, and this applies to interactions between glial cells and neurones, between various populations of neuronal cells and finally between axons and skeletal muscle fibres. The mechanisms involved in the establishment of specific synaptic connections between neurones can also be investigated in this experimental paradigm. Important information regarding this issue was also obtained on systems other than the spinal cord, i.e. the cerebellum, hippocampus and striatum. Although such information of precise connections between the host and the grafted embryonic tissue is still lacking in the spinal cord, there is much information on the response of the host nervous system to the grafted embryonic tissue, and that of the graft to its new host environment.

It appears that embryonic grafts are able to induce repair processes following injury to the nervous system. Grafting specific populations of glial or neuronal cells into the damaged spinal cord can enhance the regenerative capacity of the host. To give a few examples: glial cells that lose their ability to remyelinate demyelinated axons can be induced to do so by implants of specific populations of glial cells. Grafts of embryonic cord may act as relays between disconnected parts of the spinal cord and in this way allow the re-establishment of some function. Specific populations of neurones known to release transmitters that influence the excitability of cells in the spinal cord may be grafted so as to modify the excitability of the existing circuitry of the spinal cord of the host. The mechanism by which such influences upon the damaged nervous system are exerted are discussed in the relevant chapters. The book also considers the possibility that populations of highly specialized and unique cells such as motoneurones might be replaced by homologous cells from embryonic grafts in conditions where the host cells are lost. The ability of such grafted cells to establish afferent and efferent connections is described in detail.

The book reflects the present state of the art of the subject, where there is a great body of morphological, immunocytochemical and molecular information available as to the events that occur during graft-host interactions. However, only few well documented and thorough studies on functional consequences of the various experimental paradigms used are available. Yet without these the value of these experiments to clinical medicine is difficult to see.

In addition to the contribution of grafting experiments to our understanding of basic scientific questions, an important aspect of this publication is the attempt to relate the acquired information to relevant clinical conditions, in particular spinal cord injury and diseases affecting the spinal cord. The authors of the book feel that there is a deep rift between the interests and hence approach to the problem of basic scientists and clinicians. This lack of communication was usually considered to be due to the delay between the time new information is obtained by basic scientists and when it reaches the clinical practitioner. However, another aspect of the problem is the relative lack of understanding by basic scientists of clinical issues. This lack of understanding often encourages the uncritical belief that a particular result obtained in the laboratory is ready to be used for treatment. A much closer link to clinical practice and more direct contact with it may reduce these problems. It is hoped that the book will encourage thoughts along these lines.

The findings summarized here show that grafted tissue can survive and thrive in a host mammal, occasionally replace some lost function and re-establish a semblance of sophisticated and complex circuitries. These new insights are among the most exciting in neurobiology, for they reverse the pessimistic view about the central nervous system that claims that nothing can grow or regenerate there. This view dominated for a very long time, and results obtained with grafts give us a new vision and encourage hope that it will be possible to treat some of the incurable diseases of the CNS.

Antal Nógrádi, Urszula Slawinska and Gerta Vrbová

ACKNOWLEDGEMENTS

We wish to thank all those friends and families who provided us with support while writing this book. It was a pleasure to have the help of Debbie Bartram and András Szabó both of whom patiently put up with many of our unreasonable demands. We would also like to acknowledge the help of the Wellcome Trust, Action Research, Hungarian National Research Fund and American ALS Association for supporting our work.

Permissions to reproduce copyright material was kindly granted by Elsevier Science Publisher B.V., Federation of European Neuroscience Societies, Pergamon Press, Springer Verlag, Society for Neuroscience and John Wiley & Sons Inc. Permission to reproduce published material was given by W.F. Blakemore, M.B. Bunge, J. Guest, J.D. Houlé, K. Kalil, J. Kocsis, P. Ohara, Y. Suzuki and X.M. Xu and we would like to express our thanks to these colleagues.

CHAPTER 1

Anatomy and Physiology of the Spinal Cord

Antal Nógrádi and Gerta Vrbová

Anatomy of the Spinal Cord

Gross Anatomy

The spinal cord is part of the central nervous system (CNS), which extends caudally and is protected by the bony structures of the vertebral column. It is covered by the three membranes of the CNS, i.e., the dura mater, arachnoid and the innermost pia mater. In most adult mammals it occupies only the upper two-thirds of the vertebral canal as the growth of the bones composing the vertebral column is proportionally more rapid than that of the spinal cord. According to its rostrocaudal location the spinal cord can be divided into four parts: cervical, thoracic, lumbar and sacral, two of these are marked by an upper (cervical) and a lower (lumbar) enlargement. Alongside the median sagittal plane the anterior and the posterior median fissures divide the cord into two symmetrical portions, which are connected by the transverse anterior and posterior commissures. On either side of the cord the anterior lateral and posterior lateral fissures represent the points where the ventral and dorsal rootlets (later roots) emerge from the cord to form the spinal nerves. Unlike the brain, in the spinal cord the grey matter is surrounded by the white matter at its circumference. The white matter is conventionally divided into the dorsal, dorsolateral, lateral, ventral and ventrolateral funiculi. Each half of the spinal grey matter is crescent-shaped, although the arrangement of the grey matter and its proportion to the white matter vary at different rostrocaudal levels. The grey matter can be divided into the dorsal horn, intermediate grey, ventral horn and a centromedial region surrounding the central canal (central grey matter) The white matter gradually ceases towards the end of the spinal cord and the grey matter blends into a single mass (conus terminalis) where parallel spinal roots form the so-called cauda equina.[1]

The dorsal roots leave the dorsal horn and dorsolateral white matter, coalesce into two bundles and enter the dorsal root ganglion (DRG) in the intervertebral foramen. Immediately distal to the ganglion, the dorsal and ventral roots unite and form a trunk, the spinal nerve. The spinal nerves, which are now outside the vertebral column, converge and form plexuses and from these emerge the peripheral nerves. The number of spinal nerves and spinal segments largely corresponds to the number of vertebrae with a few exceptions: there are eight cervical, 12 thoracic, five lumbar, five sacral and one coccygeal spinal segments in humans. The number of these segments varies slightly in different species.[2]

Transplantation of Neural Tissue into the Spinal Cord, Second Edition,
edited by Antal Nógrádi. ©2006 Eurekah.com and Springer Science+Business Media.

Fine Organization of the Spinal Cord

The fine structure of the mammalian spinal cord was studied mainly on rodents, cats and primates. The most important results were those of Rexed[2,3] and Scheibel and Scheibel[4-6] on the cat spinal cord. Although the overall organization of the human spinal cord is similar to that of other mammals, there are some differences both in the cyto- and myeloarchitecture. In the past few years several studies made an effort to describe the structure of the human spinal cord and gave a detailed account of its features. Here we give a short description of the human spinal cord and where necessary refer to the important differences between human and other mammalian species (monkey, cat, rat and mice).

Cyto- and Dendroarchitecture

The laminar distribution of spinal neurons has been widely accepted. Its main advantage is its simple and comprehensive scheme of spinal cord organization and physiological properties can also be correlated to this structural arrangement.

Cytoarchitectural laminae are characterized by the density and topography of spinal neurons in the grey matter and can usually be identified on thick cross sections (Fig. 1). In addition, each lamina has its own characteristics which are particularly distinct at the level of cervical and lumbar enlargements. Most of the information about dendritic territories has been obtained by using Golgi impregnation methods. In addition to the laminar arrangement in the coronal plain, in the ventral horn the cervical and lumbar motoneurons form rostrocaudal motor columns[7] (Fig. 1).

Lamina I is the dorsalmost lamina which covers the tip of the dorsal horn. It has a loosely packed neuropil and a low neuronal density with neurons of variable size and distribution. The most typical neuron is the so-called Waldeyer[8] cell: large, fusiform neuron with disk-shaped dendritic domain.[9-12] However, in cat and rat also small and medium-sized pyramidal neurons were identified in this lamina[3,10,11,13,14] and characterised as fusiform, pyramidal and multipolar cells.[13]

Lamina II appears as a darkly stained band in Nissl-stained sections due to its high neuronal density (substantia gelatinosa, Rolando, 1824). In cat and rodents the inner and outer zones can be distinguished[2,3] although in humans there is not a clear separation between these zones. The neuronal population consists of small fusiform neurons. There are two main cell types which form the majority of the population of lamina II: the islet cells with a rostrocaudal axis and the stalked cells with a dorsoventral dendritic tree. Other types of neurons have been described such as arboreal, curly, border, vertical, filamentous and stellate cells.[9,15-18] It is possible, however, that some of these latter neurons correspond to each other or to the two main cell types. Islet cells contain GABA therefore they are considered as the inhibitory cells of this lamina.

Lamina III can be easily distinguished from lamina II by its lower neuronal density and by the presence of intermediate size neurons. This layer has a mixed population of antenna-like and radial neurons.[19-21] These cells have a simpler dendritic morphology than those in layer II.[9,22,23] Many of the above cells contain inhibitory neurotransmitters: GABA or glycine.[24]

Lamina IV in man and cat has a variety of antenna-like cells and the so-called transverse cell.[7,9,23] Most of their dendrites originate dorsally on the cell body and spread towards lamina II and III. In animals, the axons of lamina IV neurons mainly enter the spinocervical tract, which is vestigial in humans. Most probably human lamina IV neurons project to the spinothalamic tract.[25] Laterally from this lamina there is a small group of neurons embedded in the lateral funiculus: the lateral spinal nucleus[26] (Fig. 1). Its neurons project to the midbrain and brainstem and send processes to lamina IV itself.

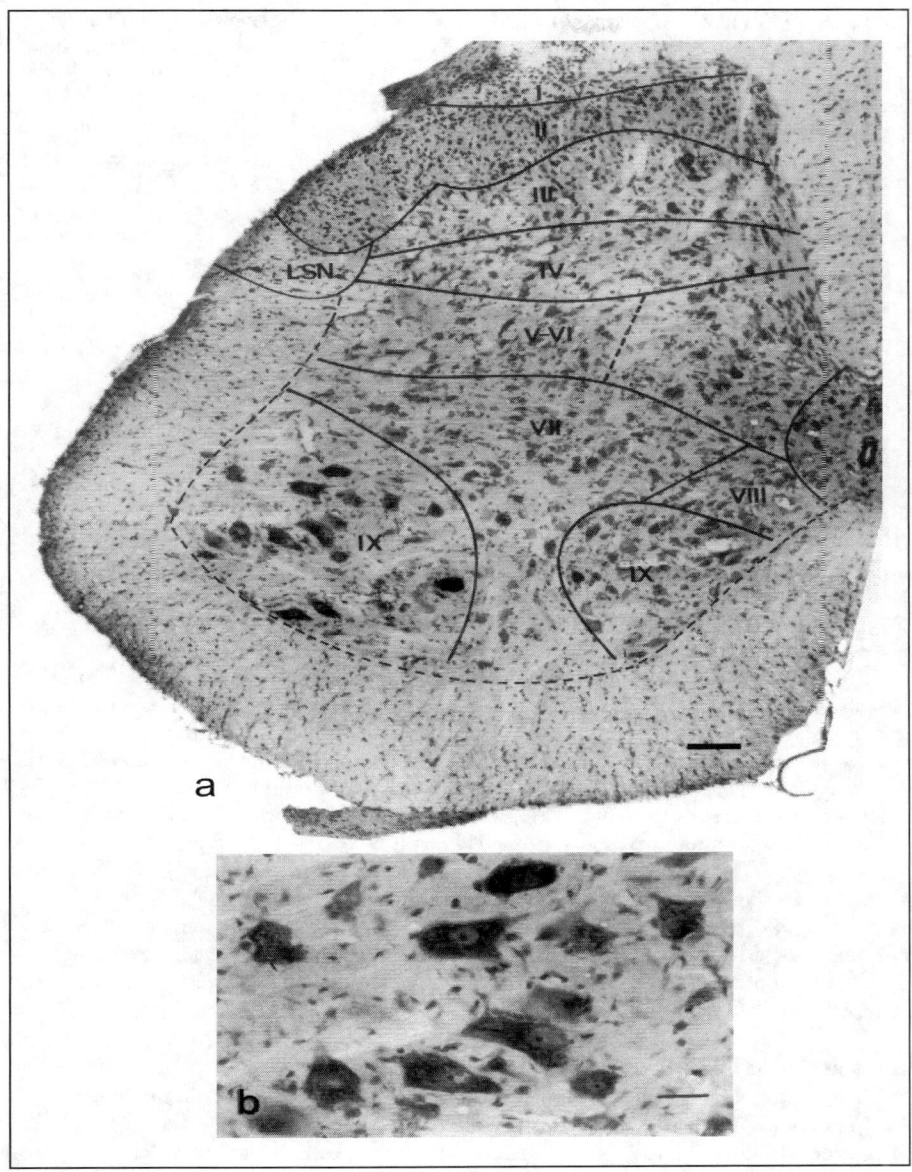

Figure 1. a) Cross section of the lumbar portion of the rat spinal cord showing the layered arrangement of the hemicord. The white matter is separated from the grey matter by a broken line. LSN= lateral spinal nucleus which takes place outside the grey matter in the dorsolateral funiculus. Cresyl violet stain. Scale bar = 50 μm. b) Higher magnification photograph shows a group of lumbar motoneurons. Note the rough and intensely stained Nissl substance which actually fills the cytoplasm. Cresyl violet stain. Scale bar = 20 μm.

Lamina V-VI have a similar cyto- and dendroarchitecture. The medial part contains fusiform and triangular neurons. The lateral part is not clearly separated from the dorsolateral funiculus. This part corresponds to the reticular formation in the brainstem and consists of medium-sized multipolar neurons.

Lamina VII occupies the intermediate zone of the grey matter and is formed by an homogeneous population of medium-sized multipolar neurons. In the appropriate segments it contains some well-defined nuclei, such as the intermediolateral nucleus (T1-L1; medially) and the dorsal nucleus of Clarke (T1-L2; laterally). The intermediolateral nucleus plays a role in the autonomic sensory and motor functions and the axons of neurons from the dorsal nucleus of Clarke form the ascending fibres of the dorsal spinocerebellar tract.

Lamina VIII has, unlike laminae I-VII a dorsoventral extension. It contains a variety of neurons with dorsoventrally polarized dendritic tree. The largest multipolar neurons can be distinguished from motoneurons only by their finer Nissl bodies by using conventional morphological techniques.[27]

Lamina IX is made up of groups of cells that form motor nuclei. Motoneurons have a unique position in this lamina, being the only spinal cord neuron which has its axon almost entirely in the peripheral nervous system. The α-motoneurons have the largest somata in the cord (50 x 70 μm) whilst the γ-motoneurons are smaller. Motoneurons can be easily recognized by the abundance of Nissl bodies in their cytoplasm and their multipolar shape (Fig. 1). Their dendrites extend for long distances, dorsally as far as lamina VI. Small neurons at the medial border of the motor nucleus are identified as the short-axoned inhibitory interneurons, the Renshaw cells. Although Rexed's classification[2,3] did not differentiate between motoneuron groups in lamina IX, these neurons can be divided into four separate columns in the human cord: the ventromedial, ventrolateral, dorsolateral and central columns[28] (Fig. 2).

Motoneurons projecting to the axial muscles are found in the ventromedial column,[28,29] those innervating proximal musculature of the limbs occupy medial and ventral position while neurons innervating distal limb muscles are located in dorsal and lateral positions. In all but one (dorsolateral) motoneuron column the dendritic polarization is longitudinal and dendritic trees overlap for a long distance (Fig. 2). Such a dendritic organization favours synchronization and synergy for axial, proximal and calf muscles.[4,5] In contrast to these columns, motoneurons in the dorsolateral column have radially oriented dendritic trees without much overlap of their dendrites (Fig. 2). This dendritic arrangement favours precise contacts with segmental afferents and may contribute to a more precise control of movements of distal muscles.

Lamina X corresponds to the substantia grisea centralis, the grey matter around the central canal. Two cell types can be recognized: (1) Bipolar cells with fan-shaped dendritic tree (dorsal portion of lamina X) and (2) bipolar cells with poorly ramified longitudinal dendrites (ventral portion).[30]

Interneurons in the Spinal Cord

Interneurons are probably the most important modulating cell types in the spinal cord. The importance of spinal interneuronal networks has only recently been acknowledged although the flexibility of these networks became apparent as early as in the 1950s.

Initially only electrophysiological approaches were used, later the precise location, morphology and immunohistochemical features helped to distinguish special interneuronal classes.

The very first morphologically and physiologically identified interneurons were the Renshaw cells and Ia interneurons (Renshaw cells project on motoneurons and thus establish the recurrent inhibition, whereas Ia interneurons are activated by Ia afferents of agonist muscles and inhibit antagonistic motoneurons).[31,32] Renshaw cells, Ia and Ib inhibitory interneurons, interneurons in disynaptic and polysynaptic reflex pathways and interneurons mediating descending commands were the "classical interneurons" and their function was thoroughly

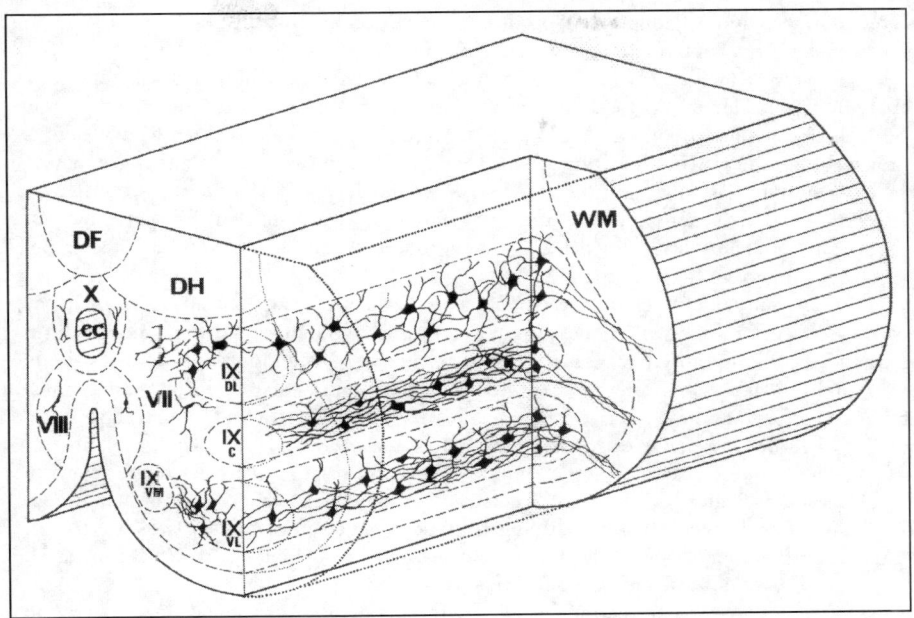

Figure 2. Schematic representation of the dendroarchitecture of spinal motoneurons in various motor columns. The ventromedial motoneurons (IX-vm) form vertical and longitudinal dendritic branches (not shown), motoneurons in the ventrolateral (IX-vl) and central (IX-c) columns tend to form dendritic bundles in the longitudinal and transverse planes. Motoneurons in these columns have long overlapping areas. On the contrary, dorsolateral motoneurons (IX-dl) have no such a dendritic bundle formation and their branches mostly branch out in the transverse plane. WM: white matter; DF= dorsal funiculus; DH= dorsal horn; CC= central canal.

analyzed in a series of studies. Recently a number of new interneurons modulating special functions were described, e.g.,interneurons involved in the clasp-knife reflex, bladder function, control of respiration and last-order premotor interneurons, etc. It is expected that the number of these highly specialized interneurons will further increase with time (for recent reviews and references see Jankowska 2001,[33] Edgley 2001[34]).

Most interneuron types have also been characterised by their neurochemical features. Renshaw cells, for example, express not only glycine, their characteristic inhibitory neurotransmitter, but they reportedly synthesize calcium binding proteins calbindin-D28k and parvalbumin.[35-37]

This short description of spinal interneurons suggests that the fine control of spinal functions mostly depends on the integrity of spinal interneuronal networks. It should be noted that interneurons named after their characteristic input (Ia, Renshaw, etc) receive a variety of multisensory inputs of different origins and these inputs together determine what the interneuron actually will do.

Glial Cells of the Spinal Cord

The central nervous system contains numerous nonneuronal, nonexcitable cells. The largest class of these cells is neuroglia or "nerve glue" a name taken from the Greek. The main glial cell types are astrocyte, oligodendrocyte, ependyma and microglia. Astrocytes together with oligodendrocytes and ependyma develop from the neuroectoderm whilst microglia is considered to be derived from blood monocytes.

Astrocytes are large cells with a stellate morphology. These very numerous fine processes radiate in all directions and contain a specific form of cytoskeletal intermediate filament, the glial fibrillary acidic protein (GFAP, Fig. 3e). Astrocytes come in two main forms: fibrous astrocytes are primarily found in white matter and protoplasmic astrocytes in the grey matter. The latter subtype has long thin processes containing much less GFAP than the fibrous astrocytes, but can be characterized by the presence of glutamine synthase. Although these types of astrocytes differ anatomically, the developmental, functional and biochemical differences between them are not fully understood.[38]

During embryonic development astrocytes guide the migration of neurons while in the mature CNS they form a structural scaffolding for other cells. Astrocytic foot processes form perivascular cuffs around CNS capillaries thus contributing to the formation of blood-brain barrier and similar processes protect the CNS from external influences at the pial surface (glial limitans externa). Apart form many other metabolic functions astrocytes are thought to transport ions and fluid from the extracellular space to vessels and they can release a number of factors which promote axonal growth.[38] Astrocytes are able react to many deleterious effect to the CNS. Morphologically this process is characterized by the appearance and proliferation of so-called reactive astrocytes (Fig. 3). Although this astrocytic healing process is often called glial repair the proliferation of astrocytes can lead to the formation of glial scar which is considered as the impediment of axonal growth and regeneration in the CNS.

Oligodendrocytes produce myelin within the CNS. One oligodendrocyte is able to myelinate several adjacent axons (Fig. 3b). The myelin is formed by these cells wrapping spiral layers of cell membrane around the axon. The inner surfaces of the cell membranes fuse and form the so-called major dense line. The myelin contains special lipids and proteins, for example the glycolipid galactocerebroside and the myelin basic protein (MBP, Fig. 3c). The myelin in the CNS is the target of several serious diseases such as multiple sclerosis and leukodystrophies. Outside the CNS myelin is formed by Schwann cells which myelinate only a single axon. Schwann cells normally are not present in the CNS (Fig. 3g) and in the case of the spinal cord and brainstem there is a distinct junction between the PNS- and CNS-type myelin called transitional zone and characterized by a complex glial structure.[39]

The CNS has its unique set of immune cells the brain macrophages. The most important and characteristic CNS macrophages are the microglial cells[40] (Fig. 3). The phenotype of microglia suggests that they are dendritic antigen-presenting cells[41] expressing class II (I-A) major histocompatibility antigens. Under pathological circumstances microglial cells become activated, increase in size and number and are usually supplemented by blood-born monocytes.

Connections of the Spinal Cord with Other Parts of the CNS

The spinal cord has its own intrinsic pathways which are called propriospinal connections. The rest of the fibre tract system connects the spinal cord to other parts of the CNS and are described here as descending and ascending pathways. There are, of course, marked species differences, the most well known are those of the corticospinal system.

Intrinsic Pathways

These tracts not only establish connections between different neuronal groups and segments of the spinal cord but also act as relays between descending pathways and intrinsic spinal neurons. Accordingly, well defined ascending and descending white matter bundles are committed to propriospinal functions.

The Lissauer's tract can be localized between the entering dorsal roots and lamina I. It is mainly composed of unmyelinated descending and ascending fibres and both types extend a

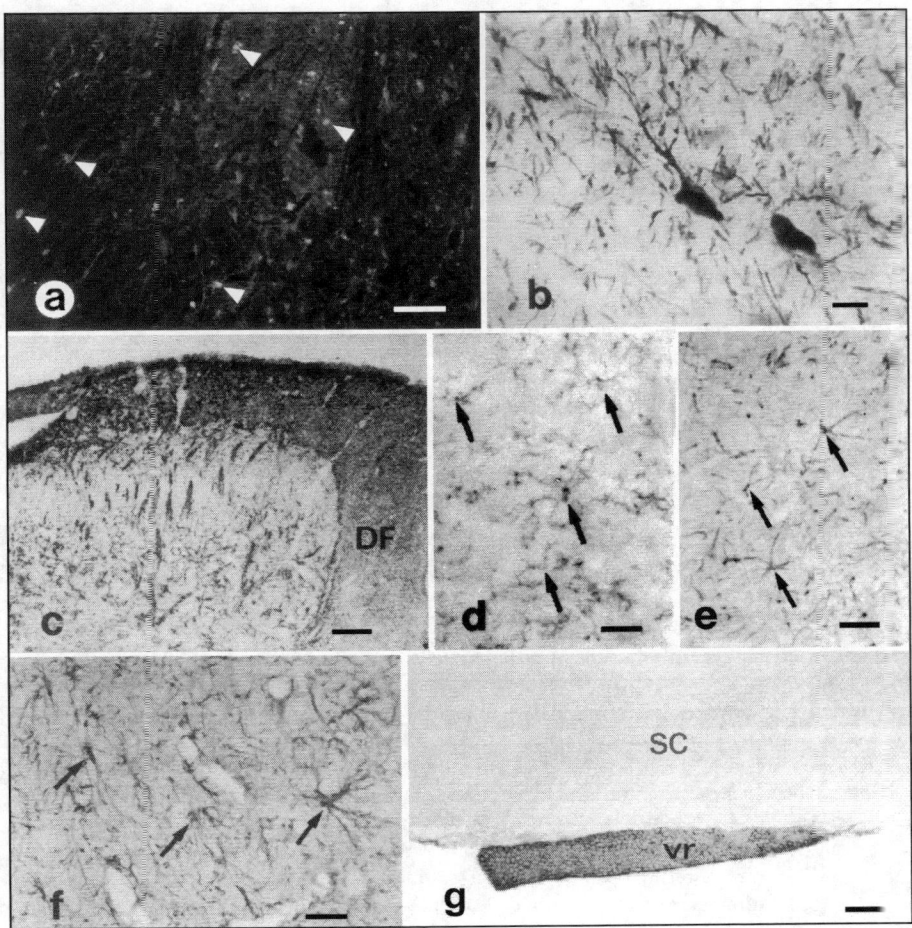

Figure 3. a) Fluorescent photograph of oligodendrocytes present in the spinal cord visualized with carbonic anhydrase II immunostaining. Scale bar = 50 μm. b) High magnification photograph shows two oligodendrocytes stained by using carbonic anhydrase enzyme histochemistry. Note the long, parallel branching processes. Scale bar = 10 μm. c) Distribution of myelinated tracts in the dorsal part of the spinal cord. Myelin sheaths are immunostained for myelin basic protein (MBP), a major protein present in normal myelin. DF: dorsal funiculus. Scale bar = 50 μm. d) Ramified microglial cells (arrows) in the grey matter of intact spinal cord. Note the faint staining and fine ramifications. Scale bar = 20 μm. Immunostaining to complement receptor type 3 (OX-42). e) Astrocytes in the intact spinal cord visualized by immunostaining to glial fibrillary acidic protein (GFAP). Scale bar = 20 μm. f) Reactive astroglial cells in an injured spinal cord (seven days after injury). Note the increased GFAP content and the thicker processes of the reactive cells (arrows). Scale bar = 20 μm. g) Low magnification photograph of the ventral part of spinal cord. The Schwann cells are immunostained with the Rat-401 antibody which is specific to Schwann cells in the adult CNS. Note that no immunostaining can be seen in the spinal cord (sc) only in the attached ventral roots (vr). Scale bar = 20 μm.

few segments. The majority of these fibres originate from the dorsal roots whilst the rest is intrinsic in nature terminating on marginal and substantia gelatinosa cells. The comma tract is a comma-shaped thin fibre bundle in between the fasciculi cuneatus and gracilis. It contains descending fibres from the cervical dorsal roots. The septomarginal tract is situated in the dorsal white matter and its position varies at the level of different segments. It consists of descending dorsal root and intrinsic fibres. The cornucommissural tract can be found along the dorsal commissure and contains ipsilaterally running descending and ascending propriospinal fibres. The anterior and lateral ground bundles are present throughout the spinal cord being most developed at the levels of enlargements. They contain both ascending and descending long and short fibres. They originate in the ipsilateral hemicord and terminate throughout the grey matter.

Ascending Pathways

The ascending pathways are formed by the central axons of dorsal root ganglion cells entering the spinal cord via the dorsal roots. They either enter an ascending fibre tract (dorsal column pathways) or terminate in the spinal grey matter. About two-third of these fibres are fine, unmyelinated, slowly-conducting C fibres. The myelinated fibre components can be classified as fast-conducting, large, myelinated Aβ, and slower-conducting, thinly myelinated Aδ fibres. Primary sensory fibres either terminate in the dorsal column nuclei of the medulla or in the superficial dorsal horn according to a segregated pattern. Thin fibres related to temperature and pain terminate in laminae I and II, whereas coarse fibres terminate in deeper layers (laminae III-V) and in the ventral horn as well (proprioceptive afferents). Furthermore, primary afferents coming from cutaneous receptors terminate almost exclusively in lamina II in rat and cat whilst visceral and muscle afferent terminals are mainly confined to laminae I and V.[42,43]

The dorsal column pathways include the medially located fasciculus gracilis (Goll) and the laterally situated fasciculus cuneatus (Burdach). The fasciculus gracilis contains dorsal root afferents from the lower limbs and lower part of the body, the fasciculus cuneatus from the upper limb and upper part of the trunk. The fibres synapse on neurons of the nucleus gracilis and nucleus cuneatus, respectively. These pathways play role in discriminative sensory tasks, such as two-point discrimination, detection of speed and direction of movements and judging of cutaneous pressure.[44] The spinothalamic tract originates from neurons in laminae I,V,VII and VIII,[25] however the distribution of spinothalamic neurons shows significant species differences. In humans the axons cross to the ventrolateral column and terminate in the ventral posterolateral and in the central lateral nuclei of the thalamus. In other mammals they terminate mainly in the posterior thalamic nuclear complex. Functionally, this tract conveys the accurate localization of pain and thermal stimuli. Ventrolateral cordotomies presented evidence that other tracts may also transmit pain stimuli.[45] The spinoreticular tract originates from cells situated bilaterally throughout the spinal grey matter. The ascending fibres in the ventral and lateral funiculi terminate in several nuclei of the reticular formation. Many spinothalamic ascending fibres also give collaterals to reticular nuclei. This pathway is responsible for carrying a variety of sensory information. The spinocervicothalamic tract uses an intermediate nucleus in the spinal cord, the lateral cervical nucleus, which is consistent in lower mammals but often absent in human spinal cords. Afferent fibres to this nucleus arise from the ipsilateral lamina IV in all cord segments. Neurons from the lateral cervical nucleus project to the contralateral thalamus via the medial lemniscus. This system is involved in tactile conditioned reflexes, tactile and proprioceptive placing and size discrimination. The spinocerebellar tracts (dorsal and ventral) carry information primarily arising from the lower extremities. The dorsal spinocerebellar tract is formed by axons of the ipsilateral nucleus dorsalis of Clarke (present in Th_1-L_2 segments in humans) and projects to the vermis and the paravermal regions of the cerebellum. It conveys information from muscle spindles, Golgi tendon organs, joints and mechanoreceptors of the lower extremities. Axons of cells situated in laminae V and VII in the lumbosacral spinal cord form the ventral spinocerebellar tract. It projects to the vermis and paravermal region of the cerebellum and probably carries information about the interrelationship of different muscle groups. Equivalent information from the upper extremities are conveyed by the cuneocerebellar and the rostral spinocerebellar tracts of the spinal cord.

Descending Pathways

The corticospinal tract is most developed in higher primates and species differences are most pronounced for this tract. The cells of origin are located in the motor cortex and their axons form the pyramidal tract. In most mammals fibres from neurons in the postcentral gyrus also contribute to this tract. In humans the bulk of the fibres cross in the lower medulla and form the lateral corticospinal tract whereas uncrossed fibres remain in the ventral funiculus and then cross in the ventral commissure. In some species the organization of this tract is different.[46,47] Functionally, the corticospinal pathway exerts a fine and amplified motor control by influencing other descending pathways.[48,49] Fibres of the reticulospinal tracts originate from the dorsal and central parts of the medulla and the pontine tegmentum. The terminal distribution of medial reticulospinal fibres is very dense in the ventral horn of the enlargements while the lateral reticulospinal tract fibres terminate in laminae I and V.[50] Fibres of the vestibulospinal tract originate from the lateral and medial vestibular nuclei. Both lateral and medial tract fibres terminate ipsilaterally in laminae VII and VII and form mono- or polysynaptic inhibitory connections with motoneurons, especially with those of neck and back muscles. The rubrospinal tract is well developed in lower mammals and less developed in humans. Its fibres originate from the caudal magnocellular part of the red nucleus and project according to a somatotopic pattern contralaterally to laminae V-VII. In cat there is a direct rubrospinal connection to motoneurons.[51] The tract exerts excitatory effects on flexor motoneurons and inhibits extensor motoneurons. The tectospinal tract tract originates from the superior colliculus and terminates contralaterally in the ventral horn of the upper cervical cord where its fibres establish multisynaptic connections with motoneurons of neck muscles.

Apart from the major descending tract, there are many minor fibre bundles originating from the interstitial nucleus of Cajal, solitary and retroambiguous nuclei, and the paraventricular nucleus of the hypothalamus. Noradrenergic fibres descend from the locus coeruleus and the lateral pontine nuclei to the grey matter and to the intermediolateral nucleus, respectively. Serotonergic projections arise from the raphe magnus and raphe pallidus and obscurus nuclei terminate either in laminae I and V (raphe magnus fibres) or in the ventral horn (rest of the fibres).

Function of the Spinal Cord

The spinal cord is a highly organized and complex part of the central nervous system Its complexity is due to the role it plays in the 3 most important functions of the individual: sensation, autonomic and motor control. If it was to simply report to the brain the information that it receives from the large number and variety of afferent inputs and relay back to the motoneurons and preganglionic neurons the outcome of processing performed by the supraspinal centres the situation would be more straight forward. However, as is well established, this is not the case and the spinal cord has, in addition to relaying information from the rest of the body to the brain and receiving efferent commands from varied portions of the brain the ability to integrate and modify both afferent signals from the periphery, and efferent signals from segmental afferents and supraspinal centres. Thus there is a complicated network of neurons that normally operates in conjunction with the rest of the CNS to allow perfect control of sensory, autonomic and motor functions. This complex circuitry is critically dependent on its connections with the brain and it can not function appropriately when it is either completely or even partially disconnected from it. It is rather regrettable that we understand so little of the potential of the complex intrinsic circuitry of the spinal cord that when it looses connection with the brain we are unable to exploit its' potential function and restore deficits caused by spinal cord lesions.

In spite of the fact that the physiology of the spinal cord has been intensively investigated for at least a century it keeps revealing new surprising phenomena.

In this chapter only a brief account will be given of its main functions.

Sensory Processing

In an oversimplified manner it can be stated that the somatic afferent functions that are processed in the spinal cord constitute the following: (a) pain and temperature, (b) touch, and (c) proprioception. Different sense organs in the peripheral structures initiate these sensory modalities, but the processing of them is usually carried out by a network of neurons in the spinal cord that are common to several of these different modalities of sensation.

Pain and Temperature

The peripheral receptors for various modalities of sensation are specialised sense organs that are contacted by axons from dorsal root ganglion neurons. These neurons have a peripheral process and a central branch that enters the spinal cord where they branch. These neurons that are directly linked with the peripheral structures are called first order neurons, and their role in processing of sensory information is largely determined by their branching pattern. Figure 4a illustrates some of the sense organs of the first order neurons that are involved in pain and temperature sensation and also shows that the main target of the branches of the central portion of this first order neuron terminates and synapses on neurons in the substantia gelatinosa. It is from this part of the dorsal horn where the second order neurons give rise to their processes which convey the information to other parts of the spinal cord and brain. However, there are ascending and descending branches of the second order neurons that synapse on cells in different segments of the spinal cord and on more ventral interneurons that are concerned with control of movement and integration of somatic afferent inputs with those from other parts of the central nervous system.[52] Thus these second order neurons play a crucial role in the processing of sensory information within the spinal cord. Not only somatic afferent fibres converge into the neurons in the substantia gelatinosa, but visceral sensation and pain also converges onto this group of second order neurons. In addition there is a strong input from various structures of the brain that impinge upon neurons in the substantia gelatinosa modify the input from the periphery and in this way the outcome of sensation (for further reading see Brown 1991,[53] Schomburg 1990[54]). It is partly because of this convergence of inputs to this part of the spinal cord that sensation is not simply the result of particular peripheral inputs.

Touch and Tactile Discrimination

The sensation of light touch is initiated from specialized sense organs in the skin or connective tissue or from free nerve endings in the dermis. The sense organs are contacted by axons from the cells of dorsal root ganglia and the information reaches the spinal cord via the central branch of the neurons of the dorsal root ganglion cells. These central branches form long tracts which give off branches to interneurons of the posterior horn in laminae VI and VII. The second order neurons within the spinal cord that process information about touch are thus in lamina VI and VII.

The same structures that are involved in the sensation of touch are also contributing to more sophisticated sensory functions such as two point discrimination, awareness of movement of body parts, as well as the position of various body parts in relation to each other. However these functions are also critically dependent on proprioception.

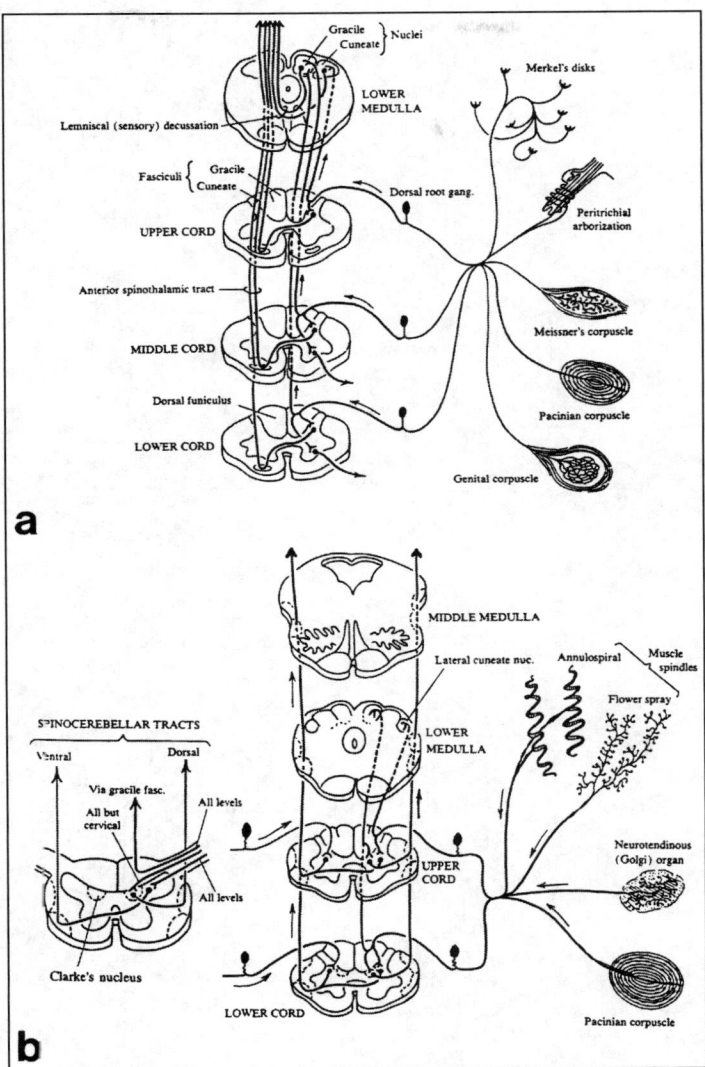

Figure 4. a) and b) illustrates schematically the types of sensory nerve endings in peripheral tissues innervated by sensory nerves (peripheral processes). It also shows the central processes and their lamination in the spinal cord and medulla. The insert on the left shows the location of the Clarke's nucleus in relation to the ventral and dorsal spinocerebellar tract and dorsal columns. Modified from ref. 88, Pansky and Allen, 1980, Review of Neuroscience.

Proprioception

The sense organs that convey this modality of sensation are located in muscles, tendons and joints (Figs. 4b and 5). The structure of these is rather complex and indicate their important function in conveying the initial signal. In the muscle the annulospiral and flower spray endings of the spindles are monitoring muscle length and this task is complicated by the fact

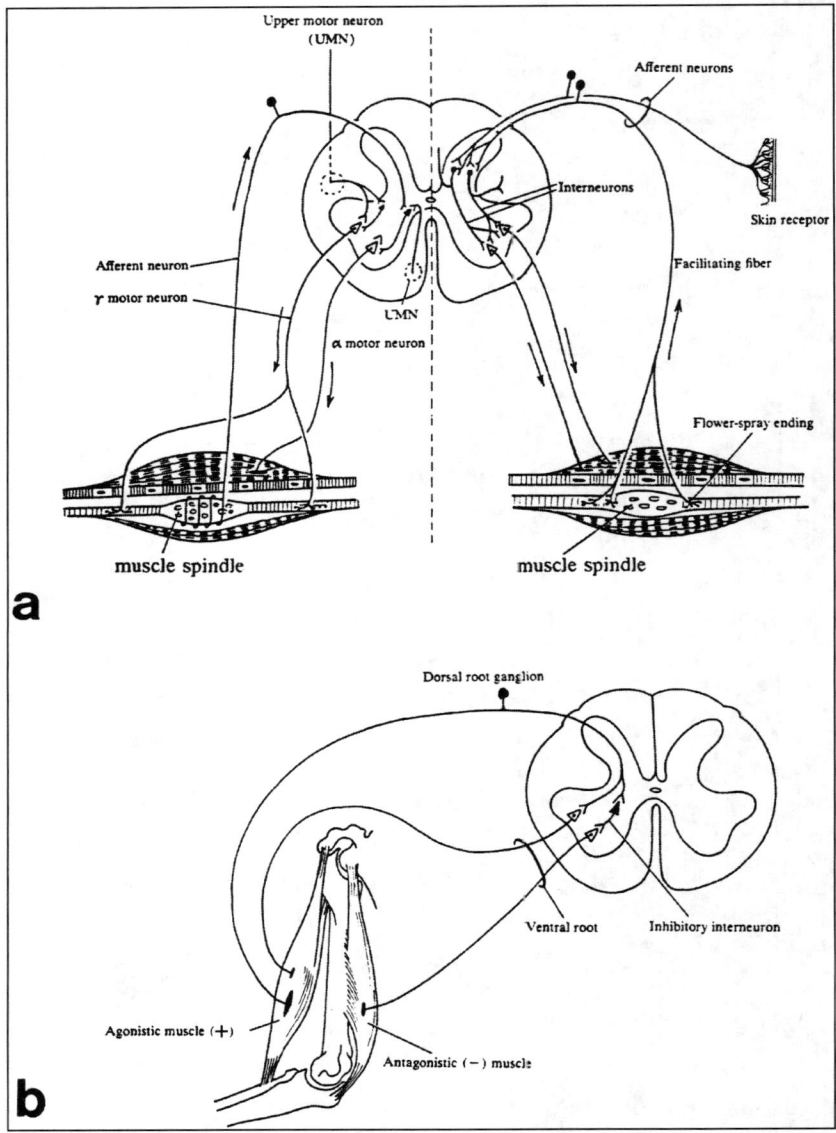

Figure 5. a) and b) The sensory and motor innervation of the mammalian muscle spindle is illustrated in a. This figure also shows the schematic picture of the cross section of the spinal cord with the various inputs to γ-motoneurons (left) and α-motoneurons (right). In b reciprocal inhibition of antagonistic muscles during a monosynaptic stretch reflex is shown. Modified from ref. 88, Pansky and Allen, 1980, Review of Neuroscience.

that the spindles are themselves a group of muscle fibres ensheathed in a connective tissue capsule and contacted by two types of sensory fibres. In addition to their sensory innervation the muscle fibres within the spindle receive their own motor innervation from small motoneurons and axons referred to as gamma efferents. Thus by relaxing or contracting the muscle fibres within the spindle the message about the state of the muscle is modified even before it reaches the spinal cord. In muscle tendons there are organs (Golgi tendon organs) which monitor the stretch imposed upon the tendon, and the Pacinian corpuscules within joints and close to bony structures monitor the pressure exerted upon these structures. The axons of sensory nerves that carry the information from the spindles towards the spinal cord are among the largest and fastest conducting nerves in the body. The central branch make up the medial division of the dorsal root as it enters the spinal cord. The central branch splits after entering the spinal cord and some of these enter the anterior horn where they synapse directly onto motoneurons to initiate a monosynaptic reflex, or onto interneurons to exert via interneurons more sophisticated control over locomotor activity.[55] These monosynaptic connections are rather unique in that there is a high degree of specificity and muscle spindle afferents from a given muscle contact only motoneurons that innervate the muscle of the origin of this afferent input. Other branches enter the posterior funiculus and ascend towards Clarke's nucleus in the posterior grey horn. Some descending and ascending branches synapse on interneurons in laminae V, VI, and VII. Axons of these cells cross the midline and ascend in the ventral spinocerebellar tract to communicate with the cerebello-olivary system.

Thus the various parts of the sensory system inform the brain about the external and internal stimuli impinging onto the extremities and trunk. However this information undergoes considerable processing by the circuitry of the spinal cord and is continuously modified by it.[54]

Motor Control

Reflexes

Our understanding of spinal cord physiology has until recently been dominated by observations of Sherrington[56] (1910) and his colleagues that the structures of the spinal cord are able to produce stereotyped responses to external stimuli. These responses were referred to as reflexes and carefully defined and observed. The simplest of these reflexes is the monosynaptic stretch reflex, elicited by activation of the IA afferent fibres that originate from the muscle spindle, and when activated produces contraction of the synonymous muscle. However even the study of this simple reflex revealed a great degree of complexity in the spinal cord circuitry. The strength of muscle contraction in response to the same stimulus was not always the same and was influenced by preceding activity of the spinal cord. In order to explain some of the findings associated with the variability of reflex activity it was necessary to consider events such as temporal and spatial summation of excitatory inputs, and inhibitory influences from other sources. Thus even the simplest "reflex" turned out to display considerable variability.[56] Nevertheless the information about the behaviour of the structures that mediate the responses to various stimuli in the spinal cord obtained by the study of reflex activity was of immense importance. It taught us that the observation of temporal and spatial summation of excitatory inputs is caused by the ability of neurons to add up excitatory postsynaptic potentials (EPSPs) and therefore when two inputs, each of which is too weak to produce a response on its own, impinge upon a neuron simultaneously, or with a slight delay, they can produce a response since the depolarisation of the cell reaches a threshold level which fires off an action potential. These rules apply even in the case of the simplest reflex response such as the stretch reflex, which is monosynaptic and the integration is carried out by only one cell, the motoneuron. All other reflexes are polysynaptic, and therefore each neuron involved in the response can contribute to the final outcome i.e., the motor response to a particular stimulus (see Fig. 5). The study of these relatively simple spinal reflexes revealed other features of the system, i.e., that neurons are not only excited, but can be inhibited by particular inputs. Such inhibition is either postsynaptic so that the membrane potential of the postsynaptic neuron increases and thus the same excitatory input fails to depolarize the neuron sufficiently to initiate an action potential, or inhibition can be presynaptic, by which the amount of excitatory transmitter released from the presynaptic terminal is reduced.

Patterned Movements Organised by the Circuitry of the Spinal Cord

In addition to the monosynaptic stretch reflex the circuitry of the spinal cord can generate patterned responses that involve movement of several joints. The best explored reflex of this type is the flexor, or withdrawal reflex in response to various sensory stimuli, and in particular in response to pain. During this reflex the extremity is withdrawn from the site of the stimulus. The flexor reflex is a complex movement which involves a highly organized sequence of activation and inhibition of motoneurons to particular muscles. It affects muscles of the contralateral limb so that the animal is supported during the time when the limb is involved in the flexor reflex and is lifted off the ground. Another patterned response that can be organized by the spinal cord is stepping. In acutely spinalised animals Brown[57] (1911) showed that the spinal cord could trigger rhythmic walking movements. These movements are of interest, since they do not depend entirely on sensory inputs and are generated by neurons located in the spinal cord. The group of neurons responsible for the organisation of this movement has been referred to as central pattern generator (CPG).[58]

Most of the information on spinal cord CPGs in mammals has been obtained on experimental animals such as rats or cats. However, whether the spinal cord of primates and humans is able to produce the same responses when disconnected from the brain is less well documented. So far the available information suggests that the isolated spinal cord of primates or humans is unable to generate such primitive stepping movements as those described for the cat.[59] Nevertheless some spinal reflex responses are preserved after complete spinal cord lesion in humans. These include the stretch reflex, which is often exaggerated and the flexor reflex. However, these responses are not stereotyped and change when they are elicited repetitively.[60] Thus even the human spinal cord is able to generate complex responses, which are influenced by repeated activity, by mechanisms that we do not understand.

The localisation of supraspinal locomotor regions is well established in the sense that electrical stimulation of such regions can elicit walking, or even galloping in decorticate cats suspended above a treadmill belt.[61] Stimulation of these areas in primates prepared in a similar manner as cats, also elicited walking and trotting. However the monkeys walked on all 4 limbs.[62] Thus like in cats the mesencephalic locomotor centre was able to activate the locomotor function of the primate spinal cord, but without the connection with this centre the stimuli that induced locomotor activity in the spinal cat were unable to do so in the spinal monkey.

Until now this section described the potential of the spinal cord to produce integrated responses without depending on the influences from the brain. However it is important to emphasize that this situation is rare and even after spinal cord injury in man the separation of the spinal cord from the brain is rarely complete. It is therefore important to consider spinal cord function in relation to the control systems that normally regulate its performance. Figure 6 summarizes the various influences from the higher centres that may influence the performance of the spinal cord circuitry.

Since this book is concerned with the possibility that neuronal or glial transplants will either replace damaged parts of the spinal cord or encourage existing structure to regenerate or resume their function, it seems pertinent to mention observations that concern the importance of various descending pathways for recovery of locomotor activity. It appears that in patients with spinal cord injuries the preservation of the ventral funiculi is best correlated with recovery of gait,[63] while patients with well preserved sensation of touch and position, but severe damage to the anterior part of the cord have a poor chance to regain the ability to walk.[59] In monkeys trained to walk on a treadmill return of locomotor performance after spinal cord injury was

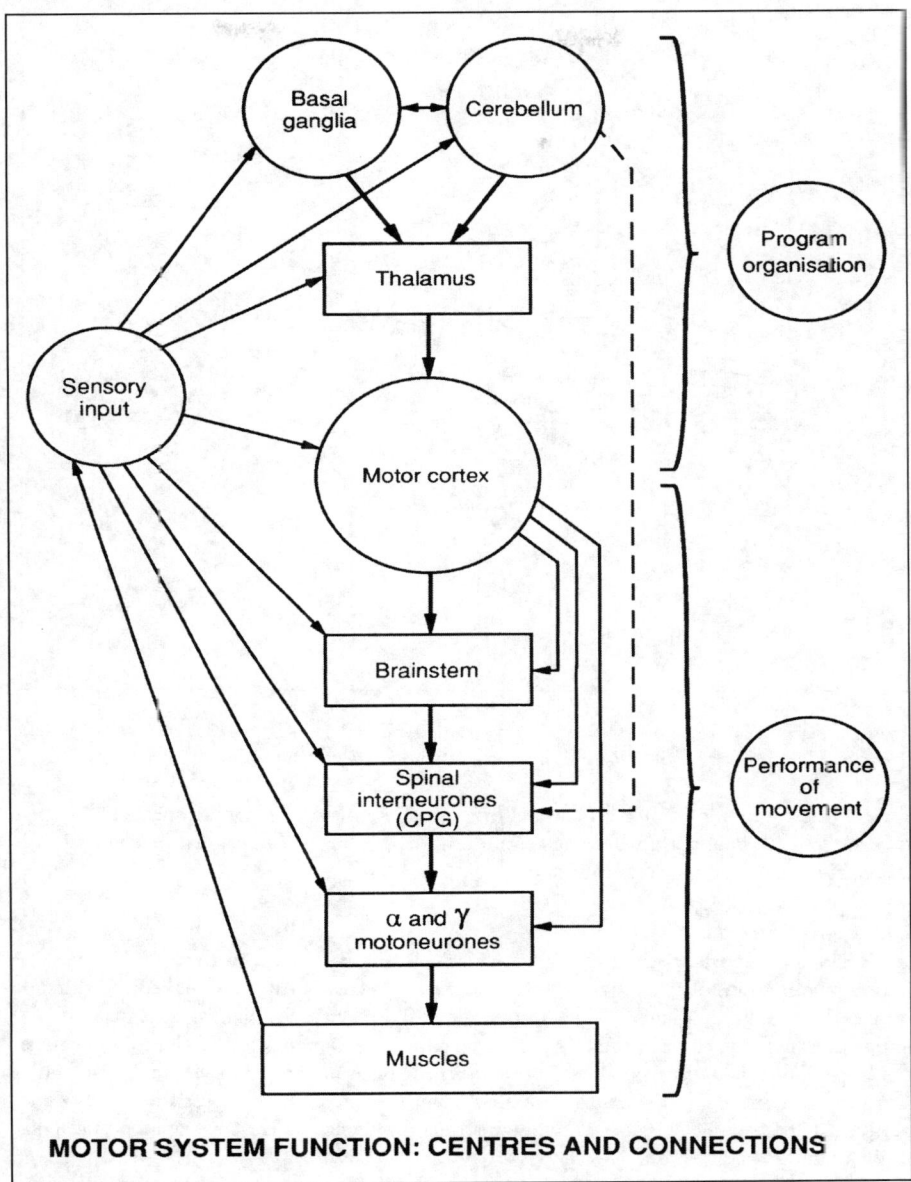

MOTOR SYSTEM FUNCTION: CENTRES AND CONNECTIONS

Figure 6. The scheme illustrates the various structures involved in the control of locomotor function.

critically dependent on the preservation of at least one ventrolateral funiculus. Retrograde la-belling of the preserved funiculus showed that the axons in the preserved funiculus originate in the vestibular, reticular and raphe nuclei.[62] Thus it appears that these structures are of critical importance for the control of the spinal cord central pattern generator. In the context of the topic discussed in this book this finding is of utmost importance, for it indicates that for

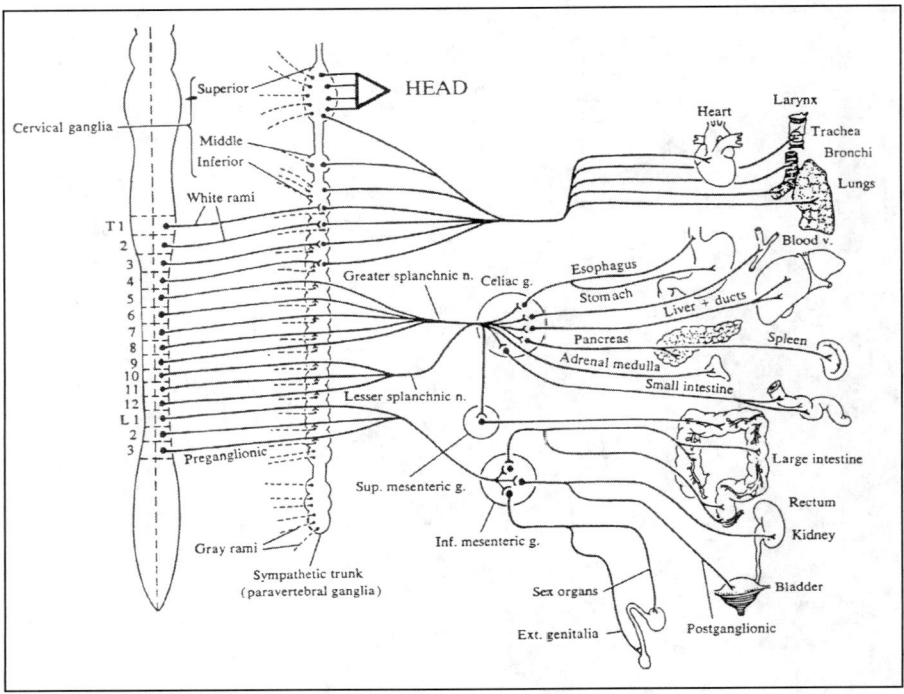

Figure 7. The figure illustrates the location of the sympathetic neurons in the spinal cord and the targets they innervate. Modified from ref. 88, Pansky and Allen, 1980, Review of Neuroscience.

successful restoration of motor function efforts should be made to reconnect particular structures, rather then haphazardly investigate indiscriminate growth of axons that may not be able to contribute to improvement of function.

Autonomic Function

There are important structures within the mammalian spinal cord that regulate various autonomic functions of the body and can be severely affected when the spinal cord is disconnected from the brain. Generally the autonomic nervous system is divided into sympathetic and parasympathetic components. The cells that control these two separate divisions occupy a typical position within the spinal cord of mammals. This is illustrated in Figure 7. The figure shows that the preganglionic neurons of the sympathetic system are localised in the thoracic and lumbar part of the spinal cord, while neurons that control the parasympathetic ganglia originate in the sacral region.

These cells that regulate important autonomic functions are closely controlled and integrated by segmental afferent inputs, and by supraspinal inputs. Following disruption of these the autonomic control of functions such as bladder control, or control of defecation as well as sexual arousal can be seriously altered and it is an important consideration that these bodily functions be restored. Much of the information on the control mechanisms exerted by the spinal cord centres over these functions is concerned with those involved in micturition. The central pathways controlling lower urinary tract function are organized as simple on-off switch

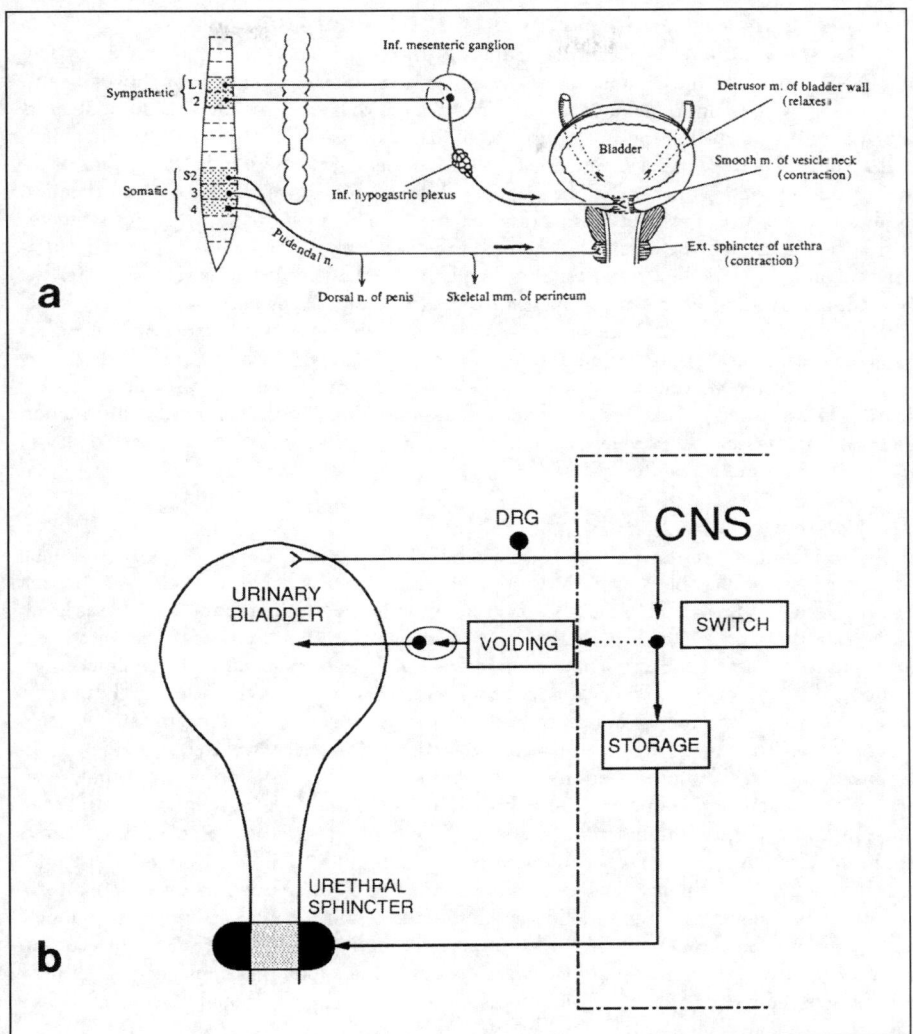

Figure 8. a) The location of the sympathetic and parasympathetic neurons in the lumbosacral cord, and the targets they innervate is shown. Modified from ref. 88, Pansky and Allen, 1980, Review of Neuroscience. b) Illustrates a scheme that could explain the control of bladder function, where the CNS can initiate the transition of the bladder from a storage to a voiding state.

circuits summarized in Figure 8. The main control is concerned with the switch from the storage of urine mode to the micturition mode.

This switching is normally accomplished by supraspinal structures, but after spinal cord injury involuntary reflex voiding can be achieved (for details see de Groat et al, 1993[64]).

Regarding other autonomic functions the information is less complete and beyond the scope of this brief summary of spinal cord physiology.

Neurotransmitters and Receptors in the Spinal Cord

The communication between the neurons of the central and peripheral nervous system and between neurons and their nonneuronal targets is established by using a variety of chemical messengers, the neurotransmitters. There are other molecules, the so-called neuromodulators which coexist with the neurotransmitters and probably regulate their function. All these molecules are different in their chemical nature as they belong to the families of amino acids, monoamines, peptides, opiates etc. The mapping of neurotransmitters involves the histochemical localization of synthesizing and degradating enzymes of the transmitter as well as recent methods, such as immunocytochemistry, receptor autoradiography, in situ hybridization, the stimulated cobalt uptake method[65] and topobiochemistry. All these recent investigations led to the better understanding of spinal mechanisms of outstanding importance, for example spinal motor functions and pain. However, it should be noted that remarkably more information is available on the mechanisms of neurotransmission in the dorsal horn and intermediate grey matter than in the ventral horn. This short overview is not intended to provide a detailed account of neurotransmitters and receptors in the spinal cord. Therefore for more information on these systems see recent reviews including those by Todd and Spike 1993,[66] Coggeshall and Carlton 1997[67] and Budai 2000.[68]

Acetylcholine

Cholinergic neurotransmission plays an outstanding role in the function of the spinal cord and therefore has been extensively studied. Although acetylcholine (Ach) was the first neurotransmitter discovered in the PNS its localization in the CNS is far not as simple as it was thought in the early 1960s. It is relatively easy to detect histochemically acetylcholinesterase, the degradative enzyme for acetylcholine but it is present in both cholinergic and cholinoceptive neurons, the latter only receiving cholinergic innervation. Cholinacetyltransferase, the enzyme synthesizing acetylcholine, is more specific for cholinergic neurons. Histochemically detected acetylcholinesterase levels as well as enzymatic levels of cholinacetyltransferase are highest in laminae I and III,[69,70] motor columns and in autonomic nuclei.[70] Nicotinic acetylcholine receptors are preferentially found in dorsal horn laminae III and IV.[71] A significant number of dorsal horn muscarinic and nicotinic receptors are thought to be located on the primary afferent terminals.[71] Muscarinic acetylcholine receptors (M1 and M2) are most abundant in laminae II and IX[72] and this fact suggests the presence of cholinergic inputs on motoneurons. Indeed, acetylcholinesterase- and cholinacetyltransferase-positive terminals were found on motor nerve cells and on Renshaw cells.[73] This important cholinergic input arises from recurrent axon collaterals from adjacent motoneurons as well as supra- and propriospinal fibres. Renshaw cells receive cholinergic afferents from motoneuron axon collaterals, though they possess nicotinic cholinergic receptors (for review see refs. 69,73,74)

In addition to the well-established roles of Ach in spinal motor performance, both muscarinic and nicotinic receptors are thought to mediate antinociceptive effects.[75]

Monoamines

The spinal cord receives an abundant monoaminergic innervation from the brainstem nuclei. Noradrenergic fibres descend from the lateral tegmentum and locus coeruleus and subcoeruleus to the dorsal and ventral horn. Serotonergic (5-HT) fibres innervate the dorsal horn (from nucl. raphe magnus) and the intermediate grey matter and ventral horn (from nucl. raphe obscurus and pallidus), whereas dopaminergic fibres from the A11 cell group of the diencephalon invade the dorsal horn (for review see: Lindvall and Björklund 1983[76]). 5-HT may be colocalized with substance P, CGRP, enkephalins and somatostatin in the raphe nuclei and their terminals.

Seven distinct 5-HT receptor subtypes have been identified ($5\text{-HT}_{1\text{-}7}$), 3 of which ($5\text{-HT}_{1\text{-}3}$) are associated with dorsal horn somatosensory processing. The activation of 5-HT receptors produce multiple physiological events as 5-HT receptors families either activate or inhibit second messenger systems.[67,68]

Dopamine D_2 receptors are mainly found in dorsal horn laminae II-III. Accordingly, focal stimulation of the A11 cell group results in selective suppression of nociceptive responses originating from multireceptive rat spinal cord neurons.[77]

There is more noradrenalin than dopamine in the spinal cord. This abundance of noradrenergic innervation is accompanied by a dense concentration of α_2-adrenergic receptors in the dorsal horn.[67,63] The clinical significance of noradrenergic neurotransmission is indicated by the finding that activation of α_2-adrenergic receptors in the dorsal horn induces analgesia in humans and experimental animals.

Amino Acids

Excitatory amino acids (EAAs), such as aspartate and glutamate are released by some interneurons (aspartate), Ia afferents and corticospinal fibres (glutamate).[78] EAAs induce their excitatory actions via two broad categories of receptors: ionotropic and metabotropic glutamate receptors. Ionotropic receptors directly regulate the opening of ion channels and three subtypes of have been distinguished: NMDA, AMPA and kainic acid (KA) receptors. Metabotropic glutamate receptors are coupled to the G-protein and their action increases the turnover of polyphosphoinositides and induces the release of intracellular Ca^{++}. EAA receptors have a relatively widespread distribution in the CNS, although some distinguished cell types display high density of certain specific receptors. Glycinergic neurons are mainly concentrated in the ventral horn: Renshaw cells and Ia interneurons are thought to use this inhibitory neurotransmitter.[79] Glycine receptors are either strychnine-sensitive or -insensitive ones. The activated strychnine-insensitive receptor is colocalized with the NMDA receptor complex and it plays a major role in the regulation of NMDA-mediated synaptic events. GABA (γ-amino butyric acid) is abundant throughout the spinal cord and GABA- as well as its synthesizing enzyme, glutamic acid decarboxylase-immunoreactive neurons are present in the ventral horn and lamina II.[80-82] $GABA_A$ and $GABA_B$ receptor subtypes have been localized on primary afferent terminals and therefore GABA is thought to participate in the presynaptic modulation of nociceptive primary afferent inputs.

Neuropeptides

A wide variety of neuropeptides is present throughout the spinal cord. The list of peptides includes somatostatin, substance P, enkephalins, calcitonin gene-related peptide (CGRP), neuropeptide Y, oxytocin, opioid peptides, nociceptin, nocistation and some others. Most of the immunoreactivity is due to fibres entering the dorsal horn of the spinal cord but also numerous various cell types contain neuropeptides.[66-68,83,84] The peptidergic immunoreactivity in dorsal horn fibres is only in part due to descending fibres from brainstem neurons, whereas many somatostatin and CGRP reactive fibres enter the dorsal horn via the dorsal root ganglia. The most intense immunoreactivity is always confined to the dorsal horn laminae where they probably play an important role in modulation of nociception.[85] Functionally, CGRP expressed by motoneurons may have a trophic action on skeletal muscle cholinergic receptors[86] but its role is obscured by the finding that some, but not all motoneurons contain this peptide. CGRP was found in most of the α-motoneurons innervating fast muscles while less motoneurons supplying slow muscles contained CGRP. In contrast, γ-motoneurons were only weakly stained for CGRP or totally devoid of CGRP labelling.[87]

References

1. Weibl H. Zur Topographie der Medulla spinalis der Albinoratte (Rattus Norvegicus). Adv Anat Embryol Cell Biol 1973; 47:6.
2. Rexed B. A cytoarchitectonic atlas of the spinal cord in the cat. J Comp Neurol 1954; 100:297-379.
3. Rexed B. The cytoarchitectonic organization of the spinal cord in the cat. J Comp Neurol 1952; 96:415-495.
4. Scheibel ME, Scheibel AB. Terminal axonal patterns in cat spinal cord. Ist ed. The lateral corticospinal tract Brain Res 1966a; 2:333-350.
5. Scheibel ME, Scheibel AB. Spinal motoneurons, interneurons and Renshaw cells. A Golgi study Arch Ital Biol 1966b; 104:328-353.
6. Scheibel ME, Scheibel AB. Terminal axonal patterns in cat spinal cord. II. The dorsal horn. Brain Res 1968; 9:32-58.
7. Brown AG. Organization in the spinal cord. The anatomy and physiology of identified neurons New York: Springer, 1981.
8. Waldeyer H. Das Gorilla Rückenmark. Abh K Akad. Berlin: Wiss, 1888:1-147.
9. Schoenen J. The dendritic organization of the human spinal cord: The dorsal horn. Neuroscience 1982a; 7:2057-2087.
10. Coimbra A, Lima D. Projections and neurochemical specificity of the different morphological types of marginal cells. In: Cervero F, Bennett GJ, Headley PM, eds. Processing of Sensory Information in the Superficial Dorsal Horn of the Spinal Cord. New York and London: Plenum Press, 1988:199-215.
11. Lima D, Coimbra A. Morphological types of spinomesencephalic neurons in the marginal zone (lamina I) of the rat spinal cord, as shown after retrograde labelling with cholera toxin subunit B. J Comp Neurol 1989; 279:327-339.
12. Lenhossék MV. Der feinere Bau des Nervensystems in Lichte neuester Forschungen. Eine allgemeine Betrachtung der Strukturprinzipien des Nervensystems, nebst einer Darstellung des feineren Baues des Rückenmarkes. Berlin: Kornfeld 1895; VII-409.
13. Lima D, Coimbra A. A Golgi study of the neuronal population of the marginal zone (lamina I) of the rat spinal cord. J Comp Neurol 1986; 244:53-71.
14. Réthelyi M, Light AR, Perl ER. Synaptic ultrastructure of functionally and morphologically characterized neurons of the superficial spinal dorsal horn of the cat. J Neuroscience 1989; 9:1846-1863.
15. Bennett GJ, Abdelmoumene M, Hayashi H et al. Physiology and morphology of substantia gelatinosa neurons intracellularly stained with horseradish peroxidase. J Comp Neurol 1980; 194:809-827.
16. Bennett GJ, Abdelmoumene M, Hayashi H et al. Spinal cord layer I neurons with axon collaterals that generate local arbors. Brain Res 1981; 209:421-426.
17. Beal JA, Nandi KN, Knight DS. Characterization of long ascending tract projection neurons and nontract neurons in the superficial dorsal horn. In: Cervero F, Bennett GJ, Headley PM, eds. Processing of Sensory Information in the Superficial Dorsal Horn of the Spinal Cord. New York and London: Plenum Press, 1988a:181-197.
18. Todd AJ, Lewis SG. The morphology of Golgi-stained neurons in lamina II of the rat spinal cord. J Anat 1986; 149:113-119.
19. Maxwell DJ, Fyffe RE, Réthelyi M. Morphological properties of physiologically characterized lamina III neurones in the cat spinal cord. Neuroscience 1983; 10:1-22.
20. Maxwell DJ. Combined light and electron microscopy of Golgi-labelled neurons in lamina III of the feline spinal cord. J Anat 1985; 141:155-169.
21. Beal JA, Russell CT, Knight DS. Morphological and developmental characterization of local-circuit neurons in lamina III of the rat spinal cord. Neurosci Lett 1988b; 86:1-5.
22. Mannen H, Sugiura Y. Reconstruction of neurons of dorsal horn proper using Golgi-stained serial sections. J Comp Neurol 1976; 168:303-312.
23. Réthelyi M, Szentágothai J. Distribution and connections of afferent fibres in the spinal cord. In: Iggo A, ed. Handbook of Sensory Physiology. Vol. II. Berlin: Springer, 1973:207-252.
24. Todd AJ, Sullivan AC. Light microscope study of the coexistence of GABA-like and glycin-like immunoreactivity in the spinal cord of the rat. J Comp Neurol 1990; 296:496-505.
25. Smith MC. Retrograde cell changes in human spinal cord after anterolateral cordotomies. Location and identification after different period of survival Adv Pain Res Ther 1976; 1:91-98.

26. Gwyn DG, Waldron HA. A nucleus in the dorsal lateral funiculus of the spinal cord of the rat. Brain Res 1968; 10:342-351.

27. Schoenen J, Faull RLM. Spinal cord: Cytoarchitectural, dendroarchitectural and myeloarchiotectura organization In: Paxinos G, ed. The Human Nervous System. San Diego: Academic Press 1990:19-53.

28. Romanes GJ. The motor columns of the spinal cord. Prog Brain Res 1964; 11:93-116.

29. Schoenen J. Dendritic organization of the human spinal cord: The motoneurons. J Comp Neurol 1982b; 211:226-247.

30. Honda C, Lee C. Immunohistochemistry of synaptic input and functional characterization of neurons near the spinal central canal. Brain Res 1985; 343:120-128.

31. Jankowska E, Lindström S. Morphological identification of Renshaw cells. Acta Physiol Scand 1971; 81:428-430.

32. Jankowska E, Lindström S. Morphology of interneurones mediating Ia reciprocal inhibition of motoneurones in the spinal cord of the cat. J Physiol 1972; 226:805-823.

33. Jankowska E. Spinal interneuronal systems: Identification, multifunctional character and reconfigurations in mammals. J Physiol 2001; 533:31-40.

34. Edgley SA. Organisation of spinal interneurone populations. J Phys 2001; 533:51-56.

35. Antal M, Freund TF, Polgár E. Calcium-binding proteins, parvalbumin- and calbindin-D 28k-immunoreactive neurons in the rat spinal cord and dorsal root ganglia: A light and electron microscopic study. J Comp Neurol 1990; 295:467-484.

36. Carr PA, Alvarez J, Leman EA et al. Calbindin-D28k expression in immunohistochemically identified Renshaw cells. Neuroreport 1998; 9:2657-2671.

37. Clowry GJ, Arnott GA, Clement-Jones M, et al. Changing pattern of expression of parvalbumin immunoreactivity during human fetal spinal cord development. J Comp Neurol 2000; 423:727-735.

38. Kimelberg HK, Norenberg MD. Astrocytes. Sci Amer April 1989; 44-52.

39. Fraher JP. The CNS-PNS transitional zone of the rat. Morphometric studies at cranial and spinal levels. Prog Neurobiol 1992; 38:261-316.

40. Jordan FL, Thomas WE. Brain macrophages: Questions of origin and interrelationship. Brain Res Rev 1988; 13:165-178.

41. Klinkert WEF. Lymphoid dendrite accessory cells of the rat. Immunol Rev 1990; 117:103-120.

42. Cervero F Dorsal horn neurons and their sensory inputs. In: Yaksh TL, ed. Spinal Afferent Processing. New York: Plenum Press, 1986:197-216.

43. Molander C, Grant G. Spinal cord projections from hindlimb muscle nerves in the rat studied by transganglionic transport of horseradish peroxidase, wheat germ agglutinin conjugated horseradish peroxidase, or horseradish peroxidase with dimethylsulfoxide. J Comp Neurol 1987; 260:246-255.

44. Giuffrida R, Rustioni A. Dorsal root ganglion neurons projecting to the dorsal column nuclei of rats. J Comp Neurol 1992; 316:206-220.

45. Willis WD, Coggeshall RE. Sensory Mechanisms of the Spinal Cord. New York: Plenum Press, 1978.

46. Molander C, Grant G. The cytoarchitectonic organization of the spinal cord in the rat. Ist ed. The lower thoracic and lumbosacral cord. J Comp Neurol 1984; 230:133-141.

47. Schoen JH. Comparative aspects of the descending fiber systems in the spinal cord. Prog Brain Res 1964; 11:203-222.

48. Kuypers HGJM. The descending pathways to the spinal cord, their anatomy and function. Prog Brain Res 1964; 11:178-200.

49. Proudlock F, Spike RC, Todd AJ. Immunocytochemical study of somatostatin, neurotensin, GABA and glycine in the rat spinal cord. J Comp Neurol 1993; 327:289-297.

50. Holstege G, Kuypers HGJM. The anatomy of brainstem pathways to the spinal cord in the cat. A labelled amino acid tracing study. Prog Brain Res 1982; 57:145-175.

51. Holstege G. Anatomical evidence for an ipsilateral rubrospinal pathway and for direct rubrospinal projections of motoneurons in the cat. Neurosci Lett 1987; 74:269-274.

52. Jankowska E, Lundberg A. Interneurones in the spinal cord. Trends Neurosci 1981; 4:230-233.

53. Brown AG. Nerve cells and nervous systems. Springer Verlag, 1991.

54. Schomburg ED. Spinal sensorimotor systems and their supraspinal control. Neurosci Res 1990; 7:265-340.

55. Jankowska E. Intraneuronal organisation in reflex pathways from proprioceptors. In: Garlik DG, Kormer PJ, eds. Frontiers in Physiol Res Australia AC of Science, 1984:228-237.
56. Sherrington CS. Flexion reflex of the limbs, crossed extension reflex and reflex stepping and standing. J Physiol 1910; 40:28-121.
57. Brown TG. The intrinsic factors in the act of progression in the mammal. Proc Roy Soc London 1911; 84:308-319.
58. Grillner S. Locomotion in vertebrates. Central mechanisms and reflex interaction. Physiol Rev 1975; 55:247-304.
59. Eidelberg E. Consequences of spinal cord lesions repair motor function, with special reference to locomotor activity. Prog in Neurobiol 1981; 17:185-202.
60. Dimitrijevic MR, Nathan PW. Studies of spasticity in man. 6. Habituation, dishabituation and sensitisation of tendon reflexes in spinal man. Brain 1973; 96:337-354.
61. Shik ML, Orlovski GN. Neurophysiology of locomotor automatism. Physiol Rev 1976; 56:465-501.
62. Eidelberg E. Locomotor control in monkeys. In: Eccles J, Dimitrijevic MR, eds. Upper Motoneuron Functions and Dysfunctions. Karger, 1985:179-184.
63. Nathan PW, Smith MC. Effects of two unilateral cordotomies on the motility of the lower limbs. Brain 1973; 96:471-494.
64. de Groat WC, Booth AM, Yoshimura N. Neurophysiology of micturition and its modification in animal models of human disease. In: Maggi CA, ed. Nervous Control of the Urogenital System, Part of series: Burnstock G, ed. The Autonomic Nervous System. Harwood Ac 1993:227-290.
65. Pruss RM, Akeson RL, Racke MM et al. Agonist-activated cobalt uptake identifies divalent cation permeable kainate receptors on neurons and glial cells. Neuron 1991; 7:509-518.
66. Todd AJ, Spike RC. The localization classical transmitters and neuropeptides within neurons in laminae I-III of the mammalian spinal dorsal horn. Prog Neurobiol 1993; 41:609-645.
67. Coggeshall RE, Carlton SM. Receptor localization in the mammalian dorsal horn and primary afferent neurons. Brain Res Rev 1997; 24:28-66.
68. Budai D. Neurotransmitters and receptors in the dorsal horn of the spinal cord. Acta Biol Szeged 2000; 44:21-38.
69. Kása P. The cholinergic systems in brain and spinal cord. Prog Neurobiol 1986; 26:211-272.
70. Aquilonius JM, Eckernas SA, Gillberg PG. Topographical localization of choline acetyltransferase within the human spinal cord and a comparison with some other species. Brain Res 1981; 211:329-340.
71. Gillberg PG, Wiksten B. Effects of spinal cord lesions and rhizotomies on cholinergic and opiate receptor binding sites in rat spinal cord. Acta Physiol Scand 1986; 126:575-582.
72. Scatton B, Dubois A, Favoy-Agid F et al. Autoradiographic localization of muscarinic cholinergic receptors at various segmental levels of the human spinal cord. Neurosci Lett 1984; 49:239-245.
73. Woolf NJ. Cholinergic systems in mammalian brain and spinal cord. Prog Neurobiol 1991; 37:475-524.
74. Krnjevic K. Transmitters in motor systems. In: Handbook of Physiology. Washington DC: Am Physiol Soc 1979:107-154.
75. Pan HL, Chen SR, Eisenach JC. Intrathecal clonidine alleviates allodynia in neuropathic rats: Interaction with spinal muscarinic and nicotinic receptors. Anesthesiol 1999; 90:509-514.
76. Lindvall O, Björklund A. Dopamine- and norepinephrine-containing neuron systems: Their anatomy in the rat brain. In: Emson PC, ed. Chemical Neuroanatomy. New York: Raven Press, 1983:229-256.
77. Fleetwood-Walker SM, Hope PJ, Mitchell R. Antinociceptive actions of descending dopaminergic tracts on cat and rat dorsal horn somatosensory neurones. J Physiol 1988; 399:335-348.
78. Urbán L, Thompson SW, Dray A. Modulation of spinal excitability: Cooperation between neurokinin and excitatory amino acid neurotransmitters. Trends Neurosci 1994; 17:432-438.
79. van den Pol AN, Görcs T. Glycine and glycine receptor immunoreactivity in brain and spinal cord. J Neurosci 1988; 8:472-492.
80. Magoul R, Onteniente B, Geffard M et al. Anatomical distribution and ultrastructural organization of the GABAergic system in the rat spinal cord. An immunocytochemical study using anti-GABA antibodies. Neuroscience 1987; 20:1001-1009.
81. McLaughlin BJ, Barber B, Saito K et al. Immunocytochemical localization of glutamate decarboxylase in rat spinal cord. J Comp Neurol 1975; 164:305-322.

82. Todd AJ, McKenzie J. GABA-immunoreactive neurones in the dorsal horn of the rat spinal cord Neuroscience 1989; 31:799-806.

83. Seybold VS, Elde RP. Immunohistochemical study of peptidergic neurons in the dorsal horn of the spinal cord. J Histochem Cytochem 1980; 28:367-370.

84. Schoenen J, Lotstra F, Vierendeels G et al. Substance P, enkephalins, somatostatin, cho_ecystoki-nin, oxytocin and vasopressin in human spinal cord. Neurology 1986; 35:881-890.

85. Yu LC, Zheng EM, Lundeberg T. Calcitonin gene-related peptide 8-37 inhibits the evoked discharge frequency of wide dynamic range neurons in the dorsal horn of the spinal cord in rat. Regul Pept 1999; 83:21-24.

86. Fontaine B, Klarsfeld A, Hökfelt T et al. Calcitonin gene-related peptide, a peptide present in spinal cord motoneurons, increases the number of acetylcholine receptors in primary cultures of chick embryo myotubes. Neurosci Lett 1986; 71:59-65.

87. Piehl F, Arvidsson U, Hökfelt T et al. Calcitonin gene-related peptide-like immunoreactivity in motoneuron pools innervating different hind limb muscles in the rat. Exp Brain Res 1993; 96:291-303.

88. Pansky B, Allen DJ. Review of Neuroscience. New York: MacMillan Publishing Co. Inc., 1980

CHAPTER 2

Recovery of Function After Spinal Cord Injury

Gavin Clowry and Urszula Slawinska

Introduction

The Factors Affecting Recovery of Function

Injuries to the spinal cord can produce variable deficits in movement, sensation and autonomic function. In humans and other primates sudden transection of the cord results
initially in a state known as "spinal shock" in which there is complete loss of sensation, in
the ability to evoke voluntary movement, and an almost complete loss of segmental reflex
responses. For complete transections the first two functions are never regained but after a period of recovery segmental reflexes return. However, these reflexes are aberrant; they recover
over a long period and change during this time. Sustained locomotor activity that leads to
stepping in response to sensory stimulation is usually difficult to elicit.

These unfortunate consequences of spinal cord injury are, however, not shared by all
vertebrates and this has given hope that information about mechanisms of recovery of spinal
cord function might be applied to mammals as well. In general it is clear that following spinal
cord injury the more primitive the species, and the younger the developmental age of the
animal, the better the recovery of function.

For instance, although there is a return to swimming in all fish, which suffer a spinal cord
transection, younger fish show better and faster restoration of function than older fish.[1] Tadpoles can restore coordinated movement whilst adult frogs do not.[2] Following transection of
the spinal cord in the newborn rat there is some recovery of function in the complete absence
of any regeneration between the two stumps of the spinal cord.[3] The rats can support themselves on their hindlimbs and develop an aberrant spinal gait. Such responses are never seen in
adult spinal rats.[4,5]

Under varying conditions, apparent recovery of function following injury to either primitive or developing central nervous system can be due to different mechanisms;

1. Particularly in the case of partial lesions there may be behavioral changes that mask the
   deficit. An animal growing up with a damaged nervous system may more easily learn strategies that allow it to utilise other parts of the nervous system not used for these purposes in
   the normal animal.
2. Recovery may have a physiological basis, removal of the influence of certain inputs may
   result in the persistence of certain preprogrammed responses, such as those of the central
   pattern generators (CPGs) which during normal development might become suppressed

Transplantation of Neural Tissue into the Spinal Cord, Second Edition,
edited by Antal Nógrádi. ©2006 Eurekah.com and Springer Science+Business Media.

and difficult to elicit segmentally. The CPG is more capable of independent functioning in lower vertebrates, and possibly in neonatal animals where it may provide the basis of some functional recovery following neonatal injury.

3. There could also be other reasons for better recovery after spinal cord injury in lower vertebrates and neonates. These may include a more favourable environment for axonal sprouting and growth. There is less glial scarring and a greater facility for axons to grow via aberrant routes to their targets. This may provide greater scope for spared pathways to occupy synaptic sites vacated by damaged pathways and, in neonates, for late growing axons to navigate lesions and still innervate the appropriate targets. In higher vertebrates though, the immediate functional effects of transection of axons more than a few millimetres from their targets for innervation is always the same; a loss of functional input from those axons which may only be compensated for but not replaced.

The direct effect of injury upon axotomized neurons often differs between adult and neonate. The injured neurone in the adult may be only able to regrow its axon for a few millimetres but the cell body and dendrites usually survive although in an atrophied state. In the neonate, axotomy often results in the death of the neurone. Therefore, lesions in the developing central nervous system often have dramatically different effects, with apparently greater sparing of function, from the same lesion as in the adult. This has been termed "the infant lesion effect". But this has to be balanced with the so-called "Gudden effect" which notes that immature neurone populations often respond to a direct lesion with a greater degree of cell death.[6] We will compare further, using specific examples as illustrations, differing types and degrees of functional restoration in response to a variety of lesions in immature and mature, and primitive and higher vertebrates.

The Effect of Spinal Cord Transection on Lower Vertebrates

Fish

The larval sea lamprey (4-5 years old but still at a developmental stage) provides both a primitive and immature vertebrate amenable to experiments in spinal cord regeneration A sufficiently rostral complete transection of the spinal cord results in paralysis caudal to the lesion. Nevertheless an apparently complete behavioral recovery of sensory and locomotor function has been observed, even after complete lesion.[7,8] Both anatomical and physiological methods have demonstrated the presence of intrinsic fibres spanning the lesion site. Two thirds of the axons of the giant reticulospinal neurons (Müller and Mauthner cells) were found to have regenerated across the scar.[9,10] Although, at 4 to 12 weeks after injury, when locomotor recovery had taken place, this growth was limited to only a few millimetres,[9,10] at 32 weeks post-transection some identifiable axons of reticulospinal neurons could be found as far as 57 mm below the transection site.[11] This axon regeneration was also confirmed both morphologically and electrophysiologically (Fig. 1). Moreover morphological studies also showed typical vesicle containing synaptic profiles in expected, as well as anomalous, regions of the cord caudal to the lesion. Müller and Mauthner axons normally make electrical synapses but these were not seen with regenerated axons.[9,12]

It is unlikely that Müller and Mauthner neurons are critical for initiation of locomotion in normal lamprey or in animals recovering from spinal cord transection,[13-15] therefore the capacity for regeneration of unidentified brain neurons, such as relatively small reticulospinal neurons that can probably activate the spinal locomotor networks and initiate locomotor behavior was also carefully investigated.[16] It was demonstrated that majority of descending brain neurons, including small, unidentified reticulospinal neurons regenerate their axons after spinal cord transection. Moreover, a greater number of projections from unidentified descending brain neurons is restored than that from Müller and Mauthner cells. Thus, these neurons

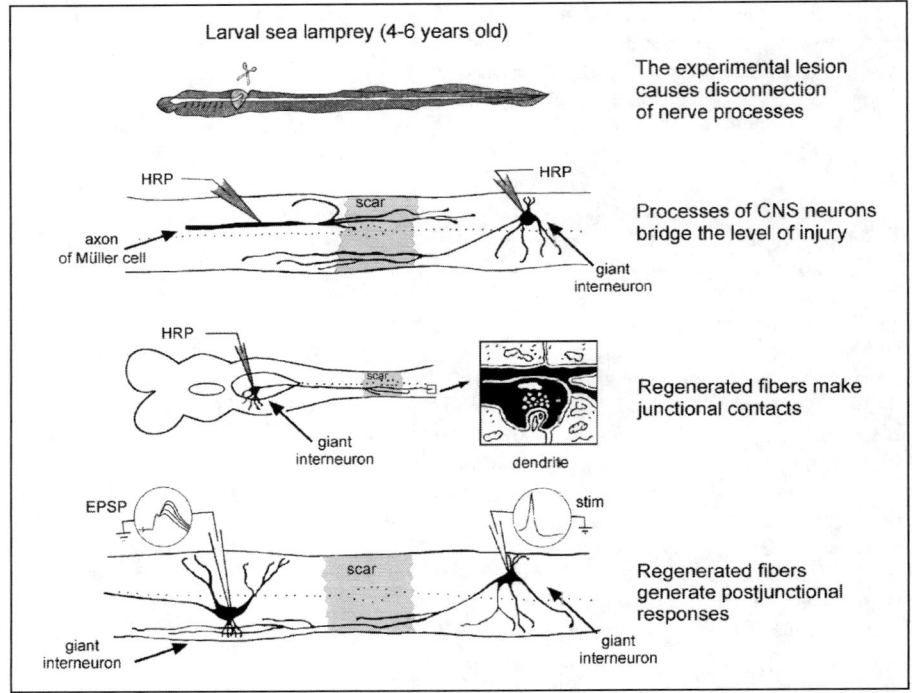

Figure 1. This figure summarises results from regeneration experiments in larval sea lampreys and illustrates how it has been demonstrated that fibres can regenerate across a spinal transection in this species and reform functional connections (modified from Cohen et al. Trends Neurosci 1988; 11:227-231).

probably contribute substantially to recovery of locomotor function in spinal cord transected lamprey since the rate of addition of new brain-spinal cord projections is relatively slow and does not appear to be enhanced after spinal cord transection.[17]

Unequivocal evidence for the regeneration of monosynaptic contacts between pairs of neurons on either side of the transection was also demonstrated. In normal cords, stimulation of the caudal interneuron elicits composite electrochemical excitatory potentials in 52% of rostral interneurons. After recovery from transection 13% of rostral neurons were recontacted. The electrical component of the input was present but reduced.[18] Similar results were found for contacts with lateral sensory cells and Müller cells.[19] Therefore, although there is an impressive degree of synaptic regeneration it falls short of reestablishing the original circuitry.

A test as to whether this axon regeneration is responsible for functional recovery involved experiments where ventral root discharges caudal and rostral to a transection were recorded.[20] The motor pattern for swimming in larval lampreys is generated by neural CPGs distributed along the spinal cord. Lengths of cord of four or more segments are capable of producing periodic ventral root discharges necessary for normal swimming movements.[21] Coordination of these discharges is possible in a cord along its entire length in the absence of descending fibres but this does require intersegmental connectivity. Similar "time-locking" of ventral root discharges is also observed in segments either side of a transection when "fictive" swimming in isolated spinal cord preparations is elicited by electrical stimulation or the application of the excitatory transmitter, glutamate. This could be a functional consequence of the reestablishment of synaptic contact between interneurons on either side of the scar. However this coordination of discharges is less efficient over longer distances than in the normal animal.[22]

The role of segmental mechanosensory feedback in the reestablishment of regenerated synaptic connections has been tested by preventing lampreys with transected cords from performing tasks they might normally reacquire by keeping them in tubular restraints. In fact, these restraints did not prevent them from regenerating swimming and crawling behaviors any less efficiently than animals recovering unrestrained.[22] These experiments demonstrate the importance of intrinsic mechanisms in the recovery of locomotor behavior and that recovery of locomotion need not be brought about by axon regeneration or synaptic rearrangement that is guided by behavioral feedback.

Although 4 weeks after spinal cord transection locomotor patterns are usually complete along the body in whole animals, in in vitro preparations deprived of mechanosensory feedback, locomotor activity was restricted to a few millimetres below the transection. Retrograde labelling studies have shown this to be the limit of regrowth of axons after this time and confirm the findings of earlier HRP intracellular labelling studies (see above). However, 32 weeks after spinal transection locomotor activity was recorded at long distances caudal to the transection in both whole and in vitro preparations. The number of brainstem neurons projecting to the rostral cord appeared the same as in normal animals whilst the projection to the caudal spinal cord, although reduced, was of a significant size.[23] Thus at longer recovery times it appears that regenerated descending axons can directly activate motor networks along the whole length of the spinal cord and initiate locomotor activity. However, it is not known to what extent the normal pattern of synaptic connectivity is restored.

For the larval lamprey, recovery of swimming behavior following spinal transection can be summarised as a functioning of CPGs free from descending control that is, to begin with, initiated and partially entrained by mechanosensory feedback from the periphery. However, the regeneration first of intersegmental axonal connectivity, and then descending pathways, seems to induce the return of a more normal pattern of locomotor activity. This further regeneration is perhaps not discernible as an improvement in the observable behavior of the animal and can only be detected in experimental situations where the spinal cord is isolated from all peripheral feedback.

Experiments in adult lampreys using partial transections showed little or no coordination of ventral root bursting either side of the lesion. However, what recovery of bursting activity there was seemed to be more reliant on regeneration of short fibre systems in the medial tract of the spinal cord rather than the regeneration of long fibres in the lateral tracts. Nevertheless these animals do recover swimming behavior after transection.[20] Similar to the injured larval lamprey soon after lesioning, the spinalised adult lamprey, too, relies on mechanosensory feedback to modulate the activity of local CPGs in the cord rather than on a system of ascending and descending fibres to coordinate activities of the CPGs. This is because it seems that regeneration even of these short fibres between segments, although it does take place, is less efficient in the adult than in the larval form. However, regeneration of descending fibres appears to play some role in the recovery in the adult with growth of both Müller and Mauthner axons below the lesion over several weeks.[24]

Whether or not regeneration of descending fibres restores the same pattern of connectivity below the lesion has been investigated in studies on goldfish. In fish 20-25 days old, transection of the spinal cord results in paralysis. Initially the fish lie on the bottom of the aquarium for up to 12 days by which time they can maintain an upright position in the water. They can then swim normally within 25 days.[25] Examination of the spinal cord 60 days after such a lesion showed that the glial scar which formed at the site of transection is not a barrier to axon regeneration[26] as is thought to be the case in the mammalian spinal cord.[27] 90% of the severed axons grew across the site of lesion but only 35-49% of the axons were able to grow as far 2cm caudal to the lesion. Interestingly, the axons able to regenerate these longer distances consisted almost entirely of large axons with diameters 4-8 times those of the average axon in the spinal cord.[28] These large axons were belonging to descending fiber tracts (tectospinal, cerebellospinal, and ventral tract group). When a second transection was made 60 days after the initial lesion this resulted again in paraplegia and silver staining revealed the presence of degenerating descending fibres.[29] Thus, the large regenerating fibres are essential for the restoration of function following spinal cord lesion even though they constitute only about a third of the normal compliment of axons. A more recent study, using retrograde labelling techniques, has shown that many descending fibres from the brainstem can, in time, regenerate the entire length of the goldfish spinal cord caudal to the transection site.[30] The remaining smaller axons make only local connections and do not regenerate for long distances.

Although this regeneration of descending axons was clearly involved in functional recovery this was not due to an exact restoration of the synaptic map in the locomotor region of the spinal cord. Using a silver stain to visualise synaptic terminals, the numbers of synapses upon intermediate grey matter neurons, ventral horn interneurons and motoneurons, 2 cm caudal to the lesion, were counted over a period of sixty days following transection of the spinal cord just rostral to the dorsal fin. Following spinal cord transection the intermediate grey matter neurons caudal to the lesion became slowly deafferented over 30 days and then remained chronically deafferented for up to 60 days. In contrast, there was a rapid deafferentation of both ventral horn interneurons and especially motoneurons following transection where about 50% of synaptic contacts were lost.[29] Therefore, under normal circumstances, large descending axons mainly contact the motoneurons directly and also, to a lesser extent, ventral horn interneurons. These connections are lost rapidly following deafferentation. These axons may not contact intermediate grey matter neurons, as the loss of synapses was much slower and may have been due to loss of polysynaptic pathways by trans-neuronal degeneration.

Over 60 days, the numbers of terminals contacting motoneurons returned to normal whilst ventral horn interneurons became hyperinnervated. However, there was not a faithful reconstruction of original pathways. Following a further spinal cord lesion, intermediate grey matter showed no further loss of terminals. Neither did motoneurons, on either the cell body or proximal dendrites. However, the number of terminals upon interneuronal somata declined by 10-15%.[29] The conclusion must be that the descending axons fail to reinnervate the motoneurons but instead innervate the ventral horn interneurons, which are probably part of the CPG. Motoneurons appeared to be reinnervated by local sprouting. This most probably derived from the same interneurons that appear to become hyperinnervated by the descending axons. Functionally, this resulted in the return of upright posture after twelve days that might be due to local sprouting reinforcing the influence of the CPG over the motoneurons. Then there was a return of coordinated swimming activity once descending axons have regenerated to the ventral horn and therefore started to influence the activity of the CPG[1] by forming contacts with the interneurons.

At one year after complete spinal cord transection the operated goldfish appeared normal in its food-catching behavior and other activities.[31,32] Using biotinylated dextran amines, which lack transcellular staining and provide clear visualization of the axonal terminals, it was demonstrated that the regenerated fibers of ascending projections followed similar pathways and terminated in similar areas of adult goldfish brain nuclei as in normal animals. Regenerated fibers had finer calibre axons and the terminal axonal arbors covered a larger area than the corresponding normal ones.[32]

All these examples indicate that in lamprey and goldfish the injured axons regenerate after spinal cord lesion without any further treatment, leading to a considerable degree of behavioral recovery. The molecular mechanisms that guide injured axons to regenerate have been studied recently on zebrafish (*Danio rerio*) and it was suggested that in this process the molecular mechanisms of axonal guidance that are present during development are also involved during regeneration.[33] It was demonstrated that axonal regrow of supraspinal neurons after spinal cord transection in adult zebrafish[34] is associated with the up-regulation of a number of growth-related genes expression.[35] To understand this phenomenon changes in the cellular environment caudal to a spinal lesion site were analysed.[33] Although a macrophage/microglial response occurred after spinal lesion, the gross cytoarchitecture of the caudal spinal cord remained intact. Moreover, myelin debris was not removed from the white matter tracts during the time of axonal regrowth. Although recovery of motor behavior after spinal cord injury in zebrafish (as in a variety of vertebrates) is likely to depend on axonal regrowth of supraspinal axons there it is still unclear to what extent the specificity of synaptic contacts of descending axons with the spinal motor system is reestablished. It was demonstrated that in this animal model the majority of descending axons enter abnormal pathways in the gray matter caudal to the lesion. It is likely that a relatively crude reconnection with the spinal intrinsic circuitry is sufficient to produce meaningful motor output after a spinal lesion, as described in other anamniotes.[36,37] Regarding regeneration in adult zebrafish it was suggested that the spontaneous regeneration of axons of supraspinal origin after spinal cord transection might be due in part to the axons' ability to negotiate novel pathways in the spinal cord grey matter.[33]

Amphibians

Amphibians can show similar powers of recovery from spinal cord transection as fish. For instance, the adult salamander is regarded as the only limbed vertebrate capable of regenerating portions of its spinal cord. Complete transections at the junction of the thoracic and lumbar spinal cord abolished all spontaneous, coordinated hindlimb and tail movements. However these returned within 60 days of the lesion. At this time, retrograde tracers introduced into the spinal cord 5-10 mm caudal to the lesion labelled up to 40% of brainstem neurons normally projecting to the lumbar cord. In addition, there was regeneration of descending axons belonging to intraspinal neurons across the transection site, which are probably involved in the coordination of fore and hindlimb movements.[38]

It is interesting that urodele amphibians are able to restore functional neuromotor units during trauma-induced tail regeneration.[39] It seems that, in urodeles, ependymal cells remain multipotential preserving the ability to differentiate into new neurons or glial cells. Thus, the repair of the CNS includes differentiation of new neurons and glial cells that become organized to form a normal spinal cord.

Perhaps of more interest is the possible effect of metamorphosis from tadpole to frog on spinal cord regeneration. There is no return of coordinated swimming or movement following spinal cord transection in young or adult frogs, nor is their evidence for regeneration of descending pathways.[2,40,41] However, there is recovery of the ability of hindlimbs to support weight and to hop in response to sensory stimulation. Tadpoles do exhibit recovery of coordinated swimming behavior following spinal cord transection. This recovery is dependent on regrowth of descending axons as it can be abolished by a second lesion.[2,41] The hindlimbs of the tadpole become very active following transection, with coordinated stepping movements observed in response to sensory stimuli. Deafferentation of the lumbar segments abolished this precocious stepping but had no effect on the recovery of swimming.[2] Intracellular recording and dye filling studies have shown that, at least, Mauthner neurons in the tadpole brainstem can regenerate axons across a spinal cord lesion and reestablish functional synaptic contact with their normal targets the spinal motoneurons.[42] However, there may not be complete return of descending control. Recovery of coordinated swimming was much better after thoracic lesions than cervical lesions, despite regrowth of descending fibres in both cases. This is because with thoracic lesions, the axial muscles provide mechano-sensory feedback to spinal cord neurons either side of the lesion which coordinates swimming movements and compensates for some loss of descending control.[43]

Furthermore, in tadpoles, regeneration of most descending fibres has been claimed to be restricted to 1-2 mm caudal to the lesion.[40] However if transection took place just prior to metamorphosis there was very extensive restoration of lumbar projections from all brainstem regions that normally send axons there. In these animals there was greater recovery of behaviors dependant on supraspinal control, such as coordinated swimming movements, than seen in lesioned tadpoles prior to metamorphosis. Retrograde labelling of neurons before and after transection showed numerous double labelled cells. This demonstrates that connections are substantially conserved through metamorphosis. Also, labelling of dividing cells showed that brainstem neurons are not generated at metamorphosis, which could subsequently project to the lumbar spinal cord.[40] Studies on different species of frog (xenopus as opposed to bullfrog) confirmed that regeneration began before metamorphosis confers behavioral recovery on juvenile frogs dependent on regeneration of descending axons not seen after metamorphosis.[41] However, this study found less convincing evidence that there is better recovery from transection during metamorphosis than seen in tadpoles prior to metamorphosis. Nevertheless, it remains an intriguing possibility that reorganisation of CNS connectivity to enable new locomotor activity appropriate to a terrestrial habitat may permit corrections in perturbations of connectivity resulting from injury.

Spinal Repair in Marsupials

Contrary to previous believes, it was confirmed many times in recent decades, that CNS axons do possess the intrinsic ability to regrow. However, adult CNS environment is not conducive to axonal regrowth after injury. The immature mammalian CNS has been considered to provide a more favourable environment for repair after injury ever since early investigations by Ramon y Cajal.[44] It was demonstrated that injury to the immature of CNS produces less irreparable damage than in the adult. It might be that the immature CNS has a greater capacity for repair because some of the growth inhibitory factors suggested to be present in the adult CNS are either absent or reduced in the developing CNS.[45] Moreover the immature neurons themselves have much greater potential for repair from injury than in the adult.[46]

The capacity for repair of an immature CNS at very early stages of development was recently investigated using the South American opossum *Monodelphis domestica*.[45,47-49] The marsupial mammal provides a unique opportunity to access an immature CNS without invasive in utero surgery, because the opossums are born at extremely immature stages of CNS development after a short gestation period of 14-15 days. Their CNS is at developmentally equivalent stage of an E13-E14 rat embryo or a 6-week-old human fetus.[50] Many spinal pathways are still developing after birth. Moreover, at birth, the neonatal opossum has unmyelinated nerve fibres that are still in the early stages of developmental growth, with only a small number of glial cells differentiated.[51] Thus, the inhibitory influences associated with myelin and its breakdown products are absent and may provide a more favourable environment for regrowth after injury.

Experiments on in vitro preparations of spinal cords from neonatal opossum had provided evidence that nerve impulse conduction across a complete spinal cord lesion returned within 4-5 days. Similar results were obtained from in vitro preparations of Monodelphis injured shortly after birth. Initially injury abolished all electrical conductivity across the lesion, but after 4 days axons showed profuse outgrowth into the lesion site followed by extensive branching as well as recovery of nerve impulse conduction.[51-53] This recovery involved growth of neurites across the lesion, some of which were regenerating from damaged axons.[54] These in vitro studies could be continued only for few days. To study longer recovery periods the in vivo opossum model were used. When the in vivo Monodelphis spinal cord was subjected to thoracic crush lesion at P4-P8 histological examination of the spinal cord immediately after the surgery showed that all axons were severed, but 10 days later numerous fibres were growing across the crushed site.[55] Moreover, the crushed spinal cord showed a predominantly normal gross spinal cord structure by 3 months of age, with numerous myelinated axons present.[49] The axons growing into a crush site preferred to grow along the pia mater, suggesting that basal lamina is a favourable substrate for the growth of fibres.[56] Moreover, even when in Monodelphis aged P4-P7 the spinal cord was transected completely, including the pia mater, creating a wider gap than caused by crushing, the histological examination performed at 2 an 6 months of age showed an intact bridge of nerve fibres across the lesion site. However, the site of spinal cord injury was still much thinner than the rostral and caudal cord. Analysis of topographical correctness of fibre growth showed that near-normal structure had been restored. Moreover, the longitudinal sections through the original crush site showed numerous nerve fibres clearly present in organized tracts.[49] It is possible that fibre growth seen following injury in the immature spinal cord is entirely the result of normal developmental growth of those ascending or descending pathways that have not reached the injury level. However, results from experiments using retrograde labelling suggest that this is not the case. Following injections of fluorescent labels below the site of injury revealed that some nuclei that have already made descending projections before the time the lesion was inflicted contained labelled cells. These were nuclei in the mid-brain central grey, dorsal medullary reticular field, lateral vestibular nucleus, raphe magnus and red nucleus. The colliculospinal and corticospinal projections had not yet reached the spinal cord by the time of injury.[49,57,58] However it is possible that following injury the originally formed neurons died due to target deprivation and some intrinsic precursor cells differentiated into neurons that grew their axons across the lesion. Whatever the mechanism the tests of motor function restoration showed remarkable development of motor behavior in Monodelphis that had received either a complete crush or cut lesion at P7-P8 including the ability to perform complex motor tasks in adulthood.[49] Animals were seen climbing a beam, crossing two grids of different sizes and swimming. However, it has not yet been determined whether the capacity for locomotion is the result of correct reinnervation formation of new sets of neurons and the growth of their axons to appropriate sites, or whether the animals have adapted to abnormal of fibre connections.[49]

The Effect of Spinal Transection on Behavior during Postnatal Development of the Rat and Kitten

There is a very clear difference in the response of the neonatal rat to complete spinal cord transection compared to the adult or weanling rat. The behavioral responses of the hindlimbs of the neonate show greater autonomy and are far less influenced by descending supraspiral influences than in the adult. Many responses seen prior to mid-thoracic spinal transection are apparent immediately after recovery from anaesthesia suggesting they are dependent on the local neuronal circuitry of the lumbar spinal cord. The duration of the spinal shock following transection of the cord is much shorter in neonates then in adults. Moreover, after neonatal spinal cord injury the development of motor responses of the hindlimbs follows a similar sequence as in normal animals, except that some responses appear earlier whilst some reflexes fail to disappear as they do in normal animals. These findings suggest that there is a repertoire of behaviors that can develop after birth even in the isolated spinal cord. They also suggest that the normal development of descending inputs serves first to inhibit and then to modify some of these reflex mechanisms. Although the responses of a neonatally transected rat never attain the fully refined characteristics of the responses of mature animals they are capable of hindlimb support, reaction to sensory stimuli, locomotor responses, and even tactile placing.[5,59] No such responses are seen following transection in adults.[4] In the adult mammal tactile placing is solely dependent upon the cerebral cortex and is permanently lost when the sensorimotor cortex is ablated.

In rats the age at which sparing of function is no longer seen after spinal cord transection is between 12 and 15 days.[4] This is a crucial time in the locomotor development of the rat. During this period rats become able to fully support their own weight during locomotion. There is elimination of polyneuronal innervation of muscle fibres. Also, there is a final matura- tion of the descending pathways of the spinal cord in the lumbar grey matter.[60] Corticospinal fibres, the final arrivals during development, penetrate the lumbar grey matter and undergo synaptogenesis.[61,62] Other descending systems such as the reticulospinal pathways, although present in the lumbosacral cord at birth, continue to undergo maturation into the third week post-natally.[63,64] It would appear that the final maturation of locomotor and neuromuscular circuitry results in modifications that allow these circuits to function only in the presence of descending, supraspinal control.

Two hypotheses exist to explain the better recovery of immature animals from spinal cord injury. Possibly in adults, the supraspinal input serves to activate intrinsic segmental circuits which lie dormant if the descending inputs are removed by transection. However, this depen- dence on descending inputs for activation never develops in the neonatally transected rat. This hypothesis has been tested by experiments in cats and will be discussed in detail below. The second hypothesis proposes that by blocking the growth of descending inputs competition between fibres of descending and segmental origin for synaptic sites is removed. This results in the persistence or formation of intrinsic circuits, which are unlike those seen in the normal animal or adult. There is evidence for this in the rat in the isolated cord of the neonatal operate where an increased projection from the dorsal root has been observed.[65,66] It has been sug- gested that ascending propriospinal axons maintain collaterals below a mid-thoracic lesion site that would be lost during the normal course of development.

Hindlimb motor behavior is not well developed in the newborn kitten and during the first few postnatal weeks overground locomotion is carried out by the forelimbs whilst the lower body makes swimming-like movements.[67] However, increased motivation such as encouraging the kitten to walk to its mother, increases hindlimb participation so that actual stepping is observed.[68] Spinal transection dramatically alters their hindlimb locomotor behavior. All overground locomotion is, of course, abolished but there is a precocious development of reflex bipedal locomotion elicited on a treadmill, which is unseen in normal neonates for several weeks.[67] Overground walking has also been observed following training and this is also the case in spinalised adult cats with weight support.[69] However, there is far more limited recovery of overground walking following transection at 12 weeks, even with training, than seen following transection 2 weeks after birth.[70] Such observations, as in rats, lead to the idea that CPGs will allow weight support and stepping in response to sensory input. However in adults this re- quires facilitation by descending inputs. In neonates, transection arrests the development of these facilitatory mechanisms but also serves to remove descending inhibition, which predomi- nates in the neonate. This allows precocious development of reflexes that are only released by increased motivation or the development of descending excitatory inputs in normally develop- ing kittens. During development, inhibitory systems within the spinal cord may take over the role of inhibition within the spinal pattern generators, as GABA blockers facilitate locomotion after spinal transection in adult cats but not in adult spinal cats following transection of their spinal cord shortly after birth. Immunocytochemistry for glutamate decarboxylase, the synthesising enzyme for the inhibitory transmitter GABA, showed reduced levels in the dorsal horn of neonatal animals compared to normal or adult operated cats.[71,72] This supports the idea that removal of descending projections in kittens alters the development of intrinsic in- hibitory synapses.

The Capacity of Spared Pathways for Sprouting Following Partial Injuries

Although regeneration of cut axons beyond the site of transection is severely limited in adult higher vertebrates,[73,74] there remains a possibility for spared axons, for instance those left undamaged by a partial section, those originating from interneurons or from other neurons located below the transection site, or sensory axons entering undamaged segments, to sprout and occupy vacated synaptic sites. Some researchers have claimed a role for such sprouting in the recovery of function after injury. However, others have doubted that a capacity for sprouting exists at all, especially in adult animals, whilst others highlight the possibility that such sprouting coupled with abnormal or inappropriate synapse formation could underlay spasticity or the emergence of chronic pain.

Experiments detecting sprouting of intact descending pathways seem to show that this response is more pronounced in neonatal animals than in adults. Partial lesions of descending pathways can lead to intact axons innervating neurons deprived of their innervation by the injury. For instance, following hemisection of the spinal cord in the rat, immunocytochemical staining reveals that the 5-HT projection caudal and ipsilateral to the lesion is 40% of that on the intact side if the lesion is made at birth, compared with a 10% level of innervation found f the same injury is made in the adult.[63] This is consistent with a superior performance in behavioral tests of reflex and locomotor function found in neonatal operates compared to adult operates.[75] A proportion of descending axons project bilaterally in the spinal cord[76] and will be spared by a hemisection. The author[69] favours the idea that, in developing animals, these axons have the ability to sprout and innervate the vacated synaptic sites. However the possibility cannot be ruled out that increased levels of innervation may be due to a failure to retract exuberant collaterals from bilaterally projecting axons in the absence of competition for synaptic sites. Another explanation is that there are late growing axons that are able to find aberrant pathways around the lesion in order to innervate their normal targets. Late growing corticospinal axons are certainly capable of such behavior (see below) as are rubrospinal axons in the opossum[77] and there is evidence that the serotonergic innervation of the dorsal horn is relatively late developing.[63]

Better evidence for such sprouting comes from studies on the development of the corticospinal tract in hamsters. Corticospinal connections are overwhelmingly crossed in this species.[78,79] However, retrograde tracing from the denervated side of the spinal cord, at various time intervals following unilateral transection of the pyramidal tract, labelled small numbers of corticospinal neurons in the ipsilateral cortex. This proves the ability of the contralateral pyramidal tract to extend new axon branches into denervated spinal cord grey matter. However, the number of such neurons able to do this declines without disappearing all together following lesioning at between 5 and 26 days postnatally. This shows that plasticity of this pathway is greater in the immature animal.

The most intensely studied projections within this context are the sensory afferents. Ever since Liu and Chambers[80] reported sprouting from an intact dorsal spinal root to innervate regions of the dorsal horn denervated by the cutting of adjacent dorsal roots. These sprouts sometimes grew over quite long distances. Other researchers have been of differing opinions as to whether, or to what extent, such sprouting occurs. No studies have confirmed growth outside the normal projection area of the intact sensory neurons. Nevertheless sprouting may occur within the normal projection area and occupy vacated synaptic sites of adjacent damaged dorsal roots. This possibility has been reported by some groups[81,82] but not confirmed by all researchers.[83-85]

However, sprouting, including some outside of the normal projection area has been reported to occur and occupy space created by the degeneration of other nearby afferent terminals in the spinal cord under two different conditions;

1. In the immature rat[86-89] undamaged afferents appear to extend into new areas of the cord they would not normally innervate. That this is genuine sprouting and not a failure to retract exuberant projections during development is confirmed by observations that retraction of exuberant collaterals is not a feature of normal sensory afferent maturation.[90-92] This type of sprouting occurs only early in the development of the animal, for instance, up to 5 days post-natally for the hindlimb afferents of the rat.[87]

2. In the adult animal following an injury to the peripheral branch of a sensory afferent.[93] Such sprouting of central afferents has been reported in "spared root" experiments[85] but also, more surprisingly, in experiments where the peripheral nerve was crushed, without accompanying dorsal rhizotomies to ensure empty synaptic sites was permanently vacated by sensory fibres, in both rats[94] and monkeys.[95] This resulted in the formation of aberrant connections and, or alternatively, the expansion of terminal fields.

Damaging the peripheral branch of a sensory nerve results in the up-regulation of the growth associated protein GAP 43, a marker for, and perhaps essential to, axon growth.[96] Following its expression it is found throughout the neurone, including the central afferents.[97] However, there is no such up-regulation following a dorsal rhizotomy. For reasons that remain unclear, a regenerative response in the periphery can also lead to a priming of new axonal growth by the central process.

Bearing in mind these findings it is worth considering the experiments of Goldberger and coworkers on partial lesions to the spinal cord of adult cats. Using a variety of behavioral tests following partial hemisection of the lower thoracic cord they found two phases of recovery. Following a 48 hour period of spinal shock there was recovery of crude locomotion and postural reflexes in the affected hindlimb. Starting two weeks after injury, accuracy of foot placement during locomotion gradually improved and postural reflexes also recovered.[69] There appeared to be a pattern to the recovery process involving a partial replacement of cortically directed fine movements with increased influence from segmental proprioreceptive inputs to the spinal CPG. This could have a physiological explanation in that it involves release of existing circuits from descending inhibition. However, the gradual onset of the phenomenon suggested an anatomical basis for recovery involving sprouting of dorsal root afferents to replace missing descending inputs ipsilateral to the section. As has been discussed above, most of the evidence suggests that such a response is unlikely without a simultaneous injury to the peripheral sensory nerve. Nevertheless, in experiments with cats, immunocytochemical staining for dorsal root afferents showed elevated staining following chronic hemisection ipsilateral to the lesion in cats in both the superficial and deep dorsal horn, the intermediate grey matter and the motor pools. GAP 43, a marker for sprouting axons,[96] was also found to be elevated ipsilateral to the lesion two weeks after hemisection.[69,98]

In parallel experiments, unilateral dorsal rhizotomy was employed to remove all sensory input from one hindlimb. This initially abolished all movement of that limb for one or two days. There was permanent abolition of all postural reflexes. However within two or three days recovery of movement was found.[99] By two weeks there is partial recovery of overground locomotion, which is dependent on descending control. This provides a clear example of behavioral substitution. Although locomotion largely recovers, the kinematic pattern of the behavior appears permanently abnormal. Quadrupedal locomotion on a treadmill, dependent on descending propriospinal control, also partially recovered.[100]

Bipedal locomotion on a treadmill, or postural control, failed to recover to any extent showing how the most basic of spinal cord mechanisms are dependent on sensory input. Also, deafferentation disrupted communication between the two sides of the spinal cord. However, there did appear to be compensation for loss of segmental inputs by increased descending control and again evidence for an anatomical basis to this recovery has been provided. Lumbosacral deafferentation of the cat spinal cord was found to lead to a synapse loss followed by a complete recovery of synapse numbers in the superficial dorsal horn and Clarke's nucleus,[101] two regions which normally receive a large sensory input. As the behavioral studies might predict, it was found that when immunocytochemical studies detecting either descending serotonergic or noradrenergic fibres, or contralateral sensory fibres, were performed there was only evidence for sprouting from descending fibres. There was complete loss of sensory fibres on the deafferented side whilst staining on the contralateral side stayed the same. Thus, there was no evidence for reinnervation of the deafferented side by the contralateral dorsal roots. In contrast there was a significant increase in serotonergic fibres in lamina II and Clarke's nucleus on the deafferented side, and probably in other areas as well.[69]

Again sprouting by descending fibres seems unlikely in light of the usual finding that such sprouting into regions of cord denuded of their own descending innervation declines with the age of the animal (see above). Nevertheless, other studies in rats support the possibility. In certain dorsal root sparing experiments, sprouting of descending serotonergic fibres led to an increased density of serotonergic innervation of Clarke's nucleus and of the superficial dorsal horn of all partially deafferented segments except the segment adjacent to the spared root. In this location, sprouting of spared sensory axons was observed.[81] This suggests that there are competitive or hierarchical mechanisms involved in determining synaptogenesis in the spinal cord. An increased density of serotonergic innervation of lamina II in the rat dorsal horn following dorsal rhizotomy has also been demonstrated by a quantitative electron microscope-immunocytochemistry study.[102] Further proof of sprouting of intrinsic spinal cord fibres in response to dorsal rhizotomy was provided by studies of substance P (SP) immunoreactivity. This neuropeptide is found in the processes of sensory fibres, descending raphe fibres and spinal cord interneurons. Dorsal rhizotomy leads to an initial fall in the density of substance P staining in laminae I and II followed by a partial return. This was interpreted as sprouting of remaining SP-ergic fibres. Electron microscopy studies showed the loss of large "scalloped" terminals, typically provided by sensory fibres, replaced by numbers of small terminals at the same synaptic sites.[103] That interneurons, as well as descending fibres, were involved in this sprouting was suggested by experiments where kainate injections into the deafferented spinal cord killed all neurons present but spared axons projecting to the region. This resulted in reduced levels of SP fibre innervation[104] presumably by killing SP containing interneurons although the possibility of transneuronal degeneration and subsequent retraction of descending SP fibres, due to the removal of their targets, cannot be discounted.

Studies on the autonomic reflexes involving ultrastructural examination of the sacral cord of the cat following transection of the thoracic cord have shown a chronic denervation of preganglionic autonomic neurons which was likely to have resulted from the loss of supraspinal inputs. However, the motoneurons of Onuf's nucleus initially lost synaptic terminals but numbers returned to normal within four days. The nature of the terminals changed. They became, on average, smaller and fewer contained pleomorphic, as opposed to round, vesicles.[105] This suggests a reorganisation of synaptic input brought about by intrasegmental sprouting. These anatomical changes would also explain the functional consequences of transection on bladder function, which include a dyssynergia between bladder voiding (under parasympathetic control) and sphincter function (elicited from Onuf's nucleus). There is also a return of reflex activity in an altered form where the micturition reflex can additionally be elicited by cutaneous stimulation. This is perhaps the result of sprouting of cutaneous afferents and associated interneurons in the spinal cord.

The Effect of Lesions to the Developing Mammalian Spinal Cord on the Neurons Innervating It

Lesions in the developing central nervous system have different effects from the same lesion in the adult. Many instances have been demonstrated of there being a greater capacity for anatomical reorganisation and sparing of function when a lesion occurs to the immature nervous system rather than in the adult. However, many immature neurons at certain stages of development will respond to injury to their axons by dying whereas such an injury to the mature neurone would not cause such a severe retrograde reaction. Pathways innervating the motoneurons and interneurons of the spinal cord undergo growth and development at different ages. At birth in rodents, dorsal root afferents are relatively mature, axons have grown into the spinal cord and reached terminal regions in the dorsal column nuclei.[65,106] Therefore spinal cord injury at birth interrupts a relatively mature sensory innervation. Some axons from brainstem spinal pathways have also reached appropriate spinal cord levels at birth but axons continue to arrive postnatally and all undergo considerable maturation and synaptic remodelling.[63,64] Their injury at birth interrupts an immature pathway.

The corticospinal pathway develops over a protracted period postnatally.[61] In rats, axons reach the level of the cervical spinal cord at 2 days postnatally and the lumbar cord by 7 days. Axons grow into the grey matter and form synapses 2 days after arriving at the appropriate level.[62] Axons from many, apparently inappropriate, parts of the cortex, for instance the visual cortex, initially project to the spinal cord but only those from the sensorimotor cortex enter the grey matter, the superfluous axons retract.[107] Spinal cord injury at birth interrupts the terrain over which corticospinal axons would grow but does not damage the axons directly. Thus, neonatal spinal cord injury affects afferent pathways at three different stages of development and allows comparisons of responses to injury at the different stages.

Response of the Dorsal Root Pathway to Injury

Cutting the peripheral process of dorsal root ganglion (DRG) neurons results in the death of 30% of axotomized neurons in adults and 75% death in neonates.[108] However, there appears to be no loss of neurons following rhizotomy of the central processes, in either adults or neonates, according to more recent reports[108,109] although this disagrees with findings of earlier research which found 50% cell death following rhizotomy of the central processes in neonates but not adults.[110]

DRG neurons react differently at the level of gene expression to peripheral axotomy as opposed to central axotomy. Only peripheral axotomy leads to up-regulation of growth associated proteins such as GAP 43.[97] This reduced ability to mount a growth response to central axotomy has been proposed as one reason why injured central processes are unable to reinnervate the dorsal horn in adult animals. In fact, central processes do grow within the peripheral portion of dorsal root populated with Schwann cells but are unable to progress beyond the interface with the CNS environment within the spinal cord.[111] Indeed, central processes will grow indefinitely if guided into lengths of peripheral nerve.[112] However, as is the case for sprouting of afferents (see above) this growth response can be boosted if a simultaneous lesion of the peripheral process is made.

It is also the case that in neonatal rats, if the central process is crushed prior to eight days post-natally, then central processes are quite able to grow back into the spinal cord. However, after eight days of age no reentry is possible.[113] Interestingly, there is no such barrier to regeneration in adult frog spinal cord.[114,115] Furthermore, dorsal root regeneration into the spinal cord can be encouraged beyond eight days post-natally in mammals. Such ingrowth can be induced in adult rat spinal cord by allowing dorsal root axons to regenerate along a substratum of embryonic astrocytes implanted into the spinal cord on millipore filters.[116]

Studies have also been carried out in which the central process of the DRG neurone has been cut at birth or 8 days post-natally, the spinal cord hemisected and a transplant of either appropriate embryonic spinal cord tissue or inappropriate embryonic cortical tissue or a gel made of extracellular matrix placed at the lesion site.[109] Two important conclusions could be inferred from the results of these experiments; firstly, that the transplants removed the barrier to sensory fibre growth normally present eight days after birth, and secondly that maturation of the regrowing fibres responded differently to different types of transplant. The cut ends of the dorsal roots were placed onto the transplant. A sub-population of DRG axons containing calcitonin gene-related peptide (CGRP) were identified immunocytochemically. Such axons grew into all three types of transplant, spinal cord, cortex or extracellular matrix following damage at either age. They remained there in excess of thirty days regardless of the type of transplant. However, patterns of innervation differed between types of transplant. In the spinal cord transplants, highly arborized plexuses of axons formed with many varicosities suggesting abundant synapse formation. In the matrigel and cortical grafts the CGRP fibres were less frequently branched and had fewer varicosities. Whilst bearing in mind that CGRP immunoreactivity is confined to a restricted population of sensory fibres, it would appear that relatively mature fibres do not require an appropriate target for innervation in order to extend their axons and perhaps even form small numbers of synapses. But an appropriate target tissue is able to encourage greater differentiation and plasticity by the regenerating sensory fibres.

Response of Descending Pathways to Injury and the Presence of Transplants in Neonates

After spinal cord lesion at birth in rodents, many immature, axotomized neurons undergo cell death.[109,117,118] Many cell populations responding in this way can be found in the brainstem including the red nucleus, locus coeruleus, raphe nuclei and lateral vestibular nucleus. Axons from all these nuclei can be found in the spinal cord grey matter at birth in rodents but these pathways continue to mature over the next three weeks of life. Taking the serotonergic fibres of the raphe nucleus as an example, these are present at all levels in the white matter by E18 and proceed to invade the grey matter over an extended period of time. This development of innervation follows a rostral to caudal gradient, and also a ventral to dorsal gradient such that, for instance, lumbar motoneurons are already innervated at birth but the lumbar superficial dorsal horn receives no innervation until 7 days post-natally and its innervation continues to mature until 21 days post-natally.[63]

Sectioning these descending pathways at birth, or soon after, results in massive cell loss in the brainstem nuclei, which occurs very rapidly. In the red nucleus, up to 40% of neurons are lost within the first 24 hours. This is in contrast to the situation in the adult animal where neurons may atrophy and lose the ability to retrogradely transport horseradish peroxidase, but little or no cell death is observed.[119,120] Therefore, these immature neurons are critically dependent on target contact for survival although this dependence is reduced to a need for contact to maintain full differentiation and function as the neurons mature, as has been described for other neuronal populations such as motoneurons,[121] sensory neurons[108] or corticospinal neurons.[122]

But how specific does target contact have to be to enable these neurons to survive? This has been investigated in experiments where a midthoracic over-hemisection was made at birth in the rat spinal cord and a transplant placed into the lesion site. Transplants of fetal spinal cord tissue supported the permanent survival of all axotomized red nucleus neurons.[118] This was compared with using grafts of fetal neocortex, hippocampus or cerebellum. Such transplants survived and grew within the spinal cord and developed morphological features characteristic of their normal development in the appropriate part of the brain. It was found that seven days post-operatively the transplants rescued red nucleus neurons from cell death but that this was not a permanent effect and by thirty days cell death had occurred in the red nucleus.[109]

These same transplants also permitted growth of brainstem serotonergic raphe axons, which can be easily identified immunocytochemically and viewed against a cresyl violet counterstain to reveal the location of the transplant. Serotonergic axons were present in all transplants, both fetal spinal cord and brain regions, 7-14 days post-operatively but only spinal cord transplants maintained the serotonergic axons permanently. These axons formed synaptic contacts within the neuropil of the graft as identified by electron microscopy. Axons were withdrawn earliest from cerebellar tissue, then from hippocampal tissue where they survived for up to 21 days and neocortex where they remained for 26 days.[109] This order reflects the relative density of serotonergic innervation in the normal rat brain suggesting these target tissues may produce differing amounts of some transmitter specific trophic factor. But clearly transmitter specificity alone is not sufficient to maintain this connectivity. Different regions of the raphe project to brain and spinal cord and it appears spinal cord projecting serotonergic neurons require contact with spinal cord tissue to thrive.

Therefore, it can be seen that brainstem axons grow into transplants, but can they grow through transplants and extend their axons into host spinal cord distal to the lesion? This was investigated in experiments where descending pathways were prelabelled with the fluorescent dye fast blue (FB) prior to making the lesion and transplant, at the same time removing all fast blue from the spinal cord. Following a survival period of 3-6 weeks another retrograde tracer diamidino yellow (DY) was injected into the spinal cord caudal to the lesion and transplant. Brainstem neurons were found that contained both FB and DY.[123] The conclusion must be that these neurons were axotomized by the lesion but regenerated axons through the transplant to innervate the regions where the DY was injected. Raphe neurons containing DY only were also found although no red nucleus neurons containing DY only were found. Such neurons presumably have late growing axons that had reached far enough down the spinal cord to be prelabelled with fast blue but arrived later to grow through the graft to more caudal regions. It is known that the serotonergic projection to the dorsal horn is relatively late in developing[63] whereas the red nucleus has reached its targets for innervation by birth in the rat. The conclusion is that transplants can facilitate plasticity by encouraging regeneration of neurons that would otherwise die and providing a bridge for late growing axons to their target sites (Fig. 2).

Individual corticospinal fibres respond in different ways to spinal cord transection in neonatal rodents and kittens depending on their exact stage of development. This fibre tract is particularly late developing. In the rat the first fibres reach the upper thoracic cord by P1, the lower thoracic segments by P3, the lumbar segments by P7 and the sacral segments by P9. However, later growing axons continue to reach the spinal cord after this time. Growth into the grey matter and synapse formation there takes place 2 days after the axons have invaded the white matter tracts of a particular segment.[61,62]

Corticospinal neurons can respond in one of three ways to a lesion of the corticospinal tract during the first two weeks of life of a rodent up to 5-6 days of age. If the axons have not yet reached the point in the tract where the lesion was made then the axons are able to find aberrant pathways by which to grow around the lesion site and continue on to innervate their normal targets. This has been demonstrated by anterograde tracing studies.[78,124-127]

Damaging axons that are still growing and had not reached their target can result in the death of their cell bodies. However, if the neuron has already made synaptic contact with neurons in the spinal cord then it responds axotomy by atrophying but it does not die. This has been demonstrated by prelabelling corticospinal neurons with fluorescent beads by retrograde transport from the spinal cord prior to transection in both kittens[125] and hamsters (Fig. 3).[122] These studies appeared to demonstrate that there is no regeneration of severed axons. The plasticity in the mammalian corticospinal pathway after early axotomy does not derive from

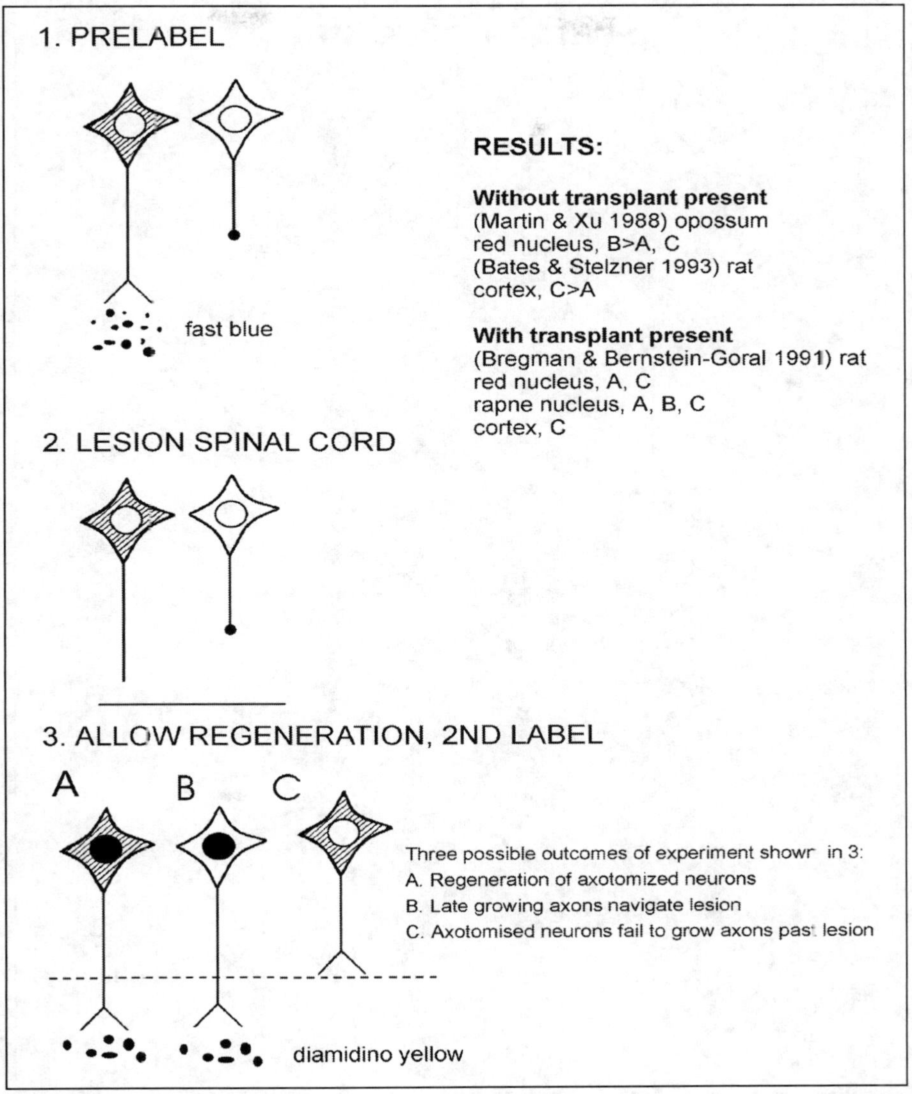

1. PRELABEL

fast blue

RESULTS:

Without transplant present
(Martin & Xu 1988) opossum
red nucleus, B>A, C
(Bates & Stelzner 1993) rat
cortex, C>A

With transplant present
(Bregman & Bernstein-Goral 1991) rat
red nucleus, A, C
rapne nucleus, A, B, C
cortex, C

2. LESION SPINAL CORD

3. ALLOW REGENERATION, 2ND LABEL

A   B   C

Three possible outcomes of experiment shown in 3:
A. Regeneration of axotomized neurons
B. Late growing axons navigate lesion
C. Axotomised neurons fail to grow axons past lesion

diamidino yellow

Figure 2. This figure summarises results from regeneration experiments in developing mammals and illustrates under what circumstances axon growth across a spinal cord transection occurs. Some regeneration of axotomized neurons has been demonstrated by some authors following lesioning but this is less likely than the death of these injured neurons. Embryonic transplants into spinal cord can both encourage regeneration across the lesion and save neurons.

injured neurons themselves but from undamaged axons. These can be late growing axons that grow around the lesion site. In addition, in experiments where only hemisection of the cord has taken place, the axons innervating the spinal cord contralateral to the injury can sprout and innervate neurons denervated by the lesion (Fig. 4).

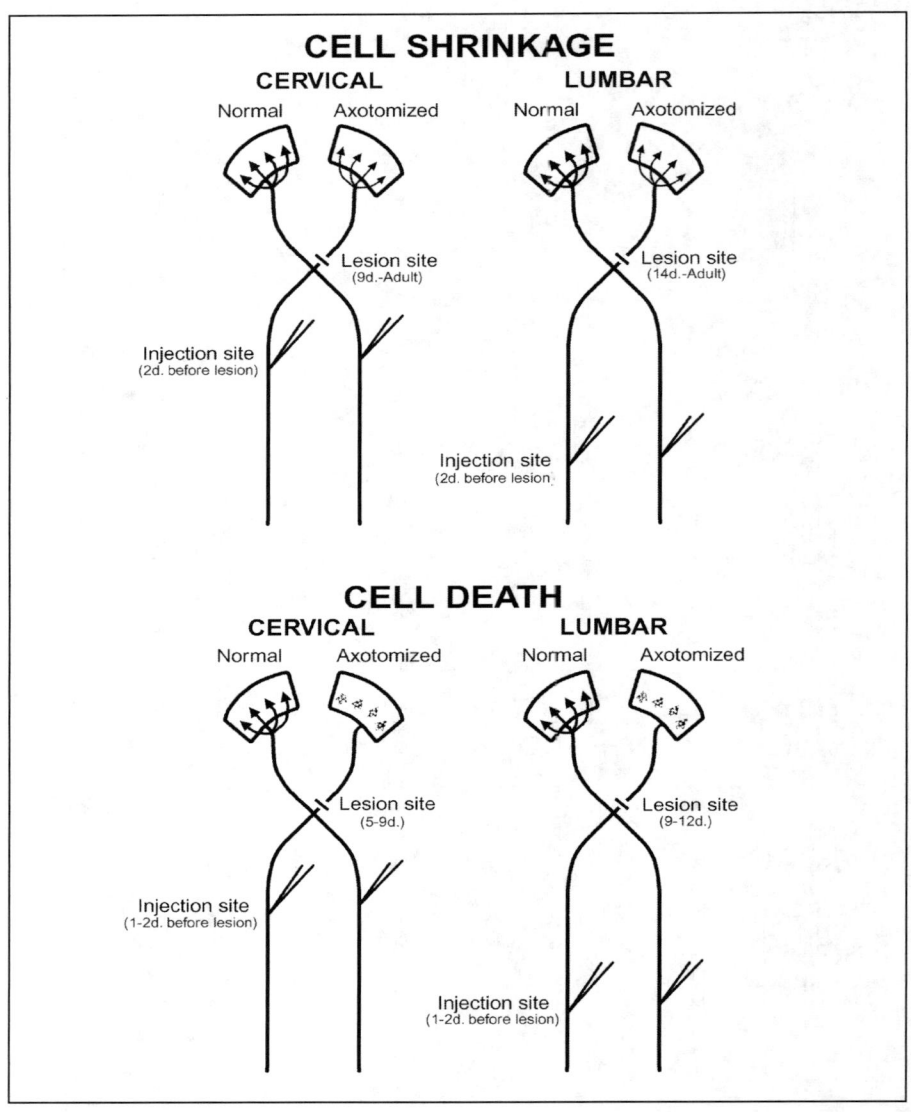

Figure 3. This figure summarises the results from an experiment on the effects of axotomy on hamster corticospinal neurons at different ages. Injections of rhodamine beads to retrogradely label corticospinal neurons were performed 1-2 days prior to a pyramidal tract lesion made rostral to the decussation. Axotomy at 9 days or later causes shrinkage of cervical projection neurons and axotomy at 14 days results in the shrinkage of the lumbar projection neurons. Cell death occurs in cervical or lumbar corticospinal populations if axotomy is performed earlier than 9 days and 14 days respectively. (Reproduced from Merline & Kalil. J Comp Neurol 1990; 296:506-516; with permission from John Wiley and Sons Inc.)

However, recent experiments using the double-labelling method of applying labels before and after lesioning found a small proportion (10-20%) of double labelled neurons, amongst those containing the second label only, showing that regeneration of severed axons had taken

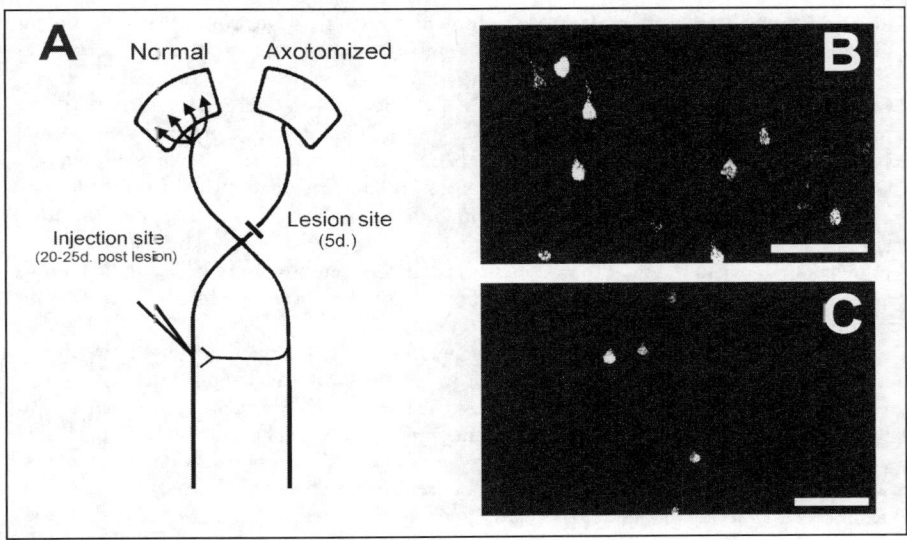

Figure 4. A) is a schematic of an experiment to demonstrate sprouting from normal corticospinal axons into denervated hamster spinal cord. The lesion site, and the site for injection of retrograde tracer, are shown. B) shows retrogradely labelled corticospinal neurons in the cervical projection area of the normal cortex as a result of sprouting after a 5 day pyramidal tract lesion, C) shows labelled neurons resulting from sprouting after a 9 day pyramidal tract lesion. Far fewer cells are labeled after the 9 day lesion demonstrating that there is less sprouting of axons as the animal matures. The scale bars = 100 μm. This figure is reproduced from Merline & Kalil. J Comp Neurol 1990; 296: 506-516, with permission from John Wiley and Sons Inc.

place.[127] this only occurred up to 6 days postnatally and was more prevalent the younger the animal. These experiments differed from earlier ones in that the lesion took place at a slightly earlier age, at a more distal lesion site and in a differing species (rat instead of hamster), which may explain the slightly differing results.

Transplants of fetal spinal cord tissue can prolong the period during which developmental plasticity of the corticospinal pathway can be elicited. Studies were made in which the mid-thoracic spinal cord was damaged at three stages in its development. Firstly, prior to the arrival of corticospinal axons, secondly, after the axons elongated through the cord but prior to synaptogenesis, and thirdly, after synaptogenesis was complete. The effects of the transplants were greatest in the first case. Anterograde tracing revealed that in all cases corticospinal fibre outgrowth was elicited but the density of synapse formation within the graft was greatest when transplantation took place prior to the arrival of corticospinal axons. However, the morphology of these labelled terminals and their synapses were similar in all cases.[128]

After spinal cord lesion and transplantation at birth, corticospinal neurons can be retrogradely labelled with injections of tracer caudal to the site of lesion and transplantation. These neurons are located in layer V of the sensorimotor cortex distributed in the restricted pattern typical of the mature animal. Therefore the transplant does not recruit and maintain the exuberant projections of pyramidal neurons from other regions of the neocortex, which are present in the neonatal rat spinal cord.[107,129] Instead, neurons undergoing a developmental plasticity in the presence of the transplant are those, which form the normal corticospinal projection of the mature animal.

The anatomical reorganisation induced by transplants of embryonic spinal cord into injured spinal cord of neonates is capable of promoting recovery of motor function. Eight weeks after hemisection of the cord, rats with a transplant crossed a grid runway more quickly and made fewer mistakes then animals without a transplant. Other quantitative and behavioral testing following behavioral training showed improvements in locomotor function.[130] It was also demonstrated that the transplantation carried out on 3-day-old rats in the site of C3 hemisection led to the reflex responses and skilled forelimb activity that resembled normal movement patterns.[131] Moreover, animals that had received a transplant developed both target reaching and accompanying postural adjustment. Further neuroanatomical investigation supported such transplant-mediated reestablishment of segmental and supraspinal input to proprospinal and motor neurons that is required for target reaching.[132] Thus, it seems that after neonatal injury in the rat, the transplant-mediated reestablisment of supraspinal input to spinal circuitry is the mechanism underlying the development of proper forelimb activity and associated postural adjustments.

Recently a correlation between axonal regeneration and functional recovery following the use of grafting technique for spinal cord repair in neonatal rats was demonstrated.[133] Using a retrograde tracing the axonal regeneration across the repaired site was quantitatively assessed in the raphe, vestibular, and red nuclei and the sensorimotor cortex. The rats that had no labelled neurons in any of the supraspinal nuclei showed no hind-forelimb coordination. The rats that had labelled neurons in the brainstem but not in sensorimotor cortex, showed hind-forelimb coordination of varying grades depending on the amount of regeneration. While the rats that had labelled neurons in all of the examined nuclei, showed almost normal locomotion. In addition to a relationship between distribution of the labelled neurons and functional recovery, a positive correlation was observed between number of labelled neurons in each of the supraspinal nuclei and locomotor performance of the rat. Thus, these results demonstrated that restored functions seem to be regulated by distribution and number of fibres regenerated across repaired site.[133]

Discussion: Possible Substrates for Recovery of Function After Injury

As we can see from the examples that have been discussed in this chapter, true regeneration of the spinal cord, that is, a complete restitution of the synaptic connections between neurons, may never occur naturally. Even in those primitive vertebrates that show the most remarkable powers of recovery, it would appear that although severed fibres are capable of regeneration, they may not grow back to reinnervate in exactly the same way the behavior generating and effecting neurons. Nevertheless, functional synaptic connections are undoubtedly made. Although it remains to be proven, perhaps the case in which greatest regeneration of normal function can be achieved is where regrowth of axons coincides with metamorphosis and the accompanying reorganisation of connectivity. Remodelling of the dendritic trees of motoneurons in metamorphosing frogs has been shown to occur[134] and this perhaps allows regenerating axons to find new synaptic sites on neurons reorganising their synaptic surface. As far as transplantation experiments are concerned, embryonic neurons transplanted into damaged spinal cord, if they continue to grow in their new surroundings, will provide new synaptic sites for ingrowing axons and so may encourage regeneration.

Possible differences between the environment of the spinal cord in lower vertebrates, which allows regeneration of transected axons, and in higher vertebrates, which does not, is considered by many researchers to lie at the root of the failure of regeneration of spinal axons in higher vertebrates rather than differences in the intrinsic ability of the neurons to regenerate axons. This is because central axons have been demonstrated to grow through peripheral nerve bridges for long distances[135] in the same way as a peripherally projecting neurone would regenerate its axon after damage to a peripheral nerve. In other words, the CNS environment of higher vertebrates does not facilitate axon growth in the same way as a peripheral nerve does. Alternatively, the CNS environment perhaps actively inhibits axon growth by mechanisms not present in peripheral nerves.

The reactive gliosis that accompanies spinal cord transection has been pinpointed as a potential barrier to regeneration[27] even though gliotic scar formation occurs in fish spinal cord through which axons are quite capable of growing.[1,22] Indeed recent evidence suggests that axons preferentially grow through the scar rather than around it.[136] This suggests inter species differences in the cells that make up the gliotic response, the main component of which is the astrocyte. It is now known that even the astrocytes of the gliotic scar of higher vertebrates fall into different categories. Reactive astrocytes secondary to a penetrative trauma (anisomorphic cells) are more permissive for axonal growth than astrocytes responding to Wallerian degeneration (isomorphic cells).[137,138] These differences seem to reside in proteins expressed on the surface membranes of the cells. However, it seems that immature astrocytes in embryonic spinal cord may actively facilitate axon growth. Cultured astrocytes from neonatal rat permitted neurite outgrowth from cultured neuronal explants[139] although type 1 astrocytes were found to be more permissive than type 2 astrocytes.[138]

The capability of axon regeneration in larval frogs and adult salamanders has been attributed to the proliferation of relatively undifferentiated ependymoglial cells[38,140] which have been observed to form a substrate for axon growth.[40,141] In juvenile frogs limited proliferation of these immature glia is associated with limited axonal regeneration, as well as the presence of something similar to astrocytic scarring.

In addition to astrocytes, another population of glia, the oligodendrocytes, appear to have a role in the inhibition of axon regeneration in adult higher vertebrates. As has already been discussed, crushed sensory afferents fail to regenerate past the PNS/CNS interface. In general CNS white matter seems to form a barrier to axon growth. It has been demonstrated that CNS myelin, produced by oligodendrocytes, contains proteins that are inhibitory to neurite growth by arresting the movements of growth cones.[142,143] In the embryonic chicken, transection of the spinal cord prior to embryonic day 13 (E13) results in complete neuroanatomical repair and functional locomotor recovery. This repair capability rapidly diminishes and is abolished between E13 and E15 coincident with myelination of fibre tracts in the spinal cord. Delaying the onset of myelination with antibodies plus homologous complement against the surface molecules of oligodendrocytes extends the permissive period for spinal cord repair.[144]

Oligodendrocyte maturation and subsequent myelination occurs post-natally in the rat spinal cord. The absence of myelin in immature nervous tissue may account for the increased capabilities for sprouting of intact axons and regeneration of sensory afferents seen in immature spinal cord. Also implicated would be the differing response of immature astrocytes more likely to be encouraging rather than inhibitory to axon growth. The aim of many transplantation experiments therefore have been to provide transplants of immature nervous tissue, or purified immature astrocytes or other axon growth promoting cells such as Schwann cells, which could facilitate the regeneration of axons otherwise blocked by the environment normally found in the mature spinal cord. Such experiments are discussed in greater detail in Chapter 4.

Therefore, although regeneration requires a permissive environment the response of the neurone is crucial; the transection of a neuron's axon may not necessarily result in a prolonged regenerative response but instead cause a limited response, such as that seen in response to the crushing of a sensory afferent. A far worse outcome is the death of the axotomized neuron, as is seen for many developing neurons (see above). A regenerative response might be possible if these neurons could be prevented from dying. For instance, in adult rats, red nucleus fibres transected in the thoracic cord sprout above the site of transection although they are unable to cross the site of transection to reinnervate their normal targets.[145] Such an injury in young rats may result in a more vigorous sprouting response but only from those few neurons that survive the injury. As we have already seen, embryonic neural transplants can save neurons from cell death as well as promote regeneration of their axons but what is the mechanism that underlies this?

One possible explanation is that growing axons are dependent with interactions with the environment they encounter on the way to the their targets. Possible substrate could include the radial glia and other astrocytes, the extracellular matrix, or other axons. As well as providing mechanisms for axon guidance, both astrocytes[146] and extracellular matrix molecules[147] have been demonstrated to have a trophic effect upon certain classes of neurons in vitro. A possible mechanism for this has been suggested from work on regenerating peripheral nerves. In mature animals, axotomy of peripheral nerves induces expression of both neurotrophins[148,149] and their receptors[150] on the surfaces of Schwann cells forming the bands of Büngner into which the regenerating axons grow. The regeneration of axons coincides temporally and spatially with a decrease in receptor expression.[150] It is proposed that Schwann cells bind neurotrophic factors to their surface forming a substratum laden with both trophic support and chemotactic guidance. No such response is seen in mature astrocytes or oligodendrocytes in response to axotomy. This may provide another explanation for failure of regeneration in the mature spinal cord. However, it is not known whether or not such mechanisms exist in the immature CNS, but transplants may save cells by providing trophic support in this fashion.

Experiments described in this chapter showed that embryonic spinal cord transplants rescued neurons more effectively than transplants from other regions of the CNS. It is possible that the extracellular matrix molecules, and cell surface proteins expressed by glial cells differ between spinal cord and the various regions of the brain. However, the most striking difference between the transplants is in the neuronal make up of the grafts. In light of the evidence that synaptic contact is made between host fibres and neurons in the transplants, the influence of the grafts on host descending or sensory fibres can be thought in one of two ways:

1. As a source of available synaptic sites with which the fibres are able to interact functionally.
2. As a source of target derived neurotrophic factors.

The second option is most usually put forward by researchers as an explanation for their findings. In its original formulation, neurotrophic factor theory was applied to peripherally projecting neurons such as sensory neurons, sympathetic neurons and motoneurons. It proposed that neurons were dependent for survival upon target derived growth factors during certain stages of development.[151] Therefore, it has been suggested that transplants may provide new sources of target derived trophic factors inaccessible to injured neurons following transection and so promote their survival. However, the evidence does not suggest that neurons providing descending innervation of the spinal cord are at any time target dependent for survival. Synaptic contact with their targets renders them immune to cell death by axotomy although such a lesion can lead to neuronal atrophy and a reduction in metabolic activity.[120,122] Projection neurons from the brain are only prone to die if their axons are cut whilst they are growing towards their target. If there is any target dependence for survival it must therefore be acting at long range and be mediated by diffusible substances. Evidence for the existence of such diffusible factors is scant although experiments with cultured embryonic tissue show growth of neurites from corticospinal neurons towards explants of cervical spinal grey matter.[152] This suggests at least a short range tropic influence may be mediated by diffusible molecules which may be involved in sprouting of axons into grafts or even towards denervated neurons resulting from partial lesions. However, there need not necessarily be any trophic effect involved.

The finding that host fibres form and maintain synaptic contact draws less comment except that it offers hope that neurons in the graft can form relays across the transection site. It does not take into account the finding that, in the case of the neuromuscular junction, synaptic contact without electrophysiological activity is ultimately as detrimental to the motoneurone as axotomy.[120,153] Descending fibres innervating spinal grafts may become incorporated into active neuronal circuitry, involving grafted neurons, whose patterns of activity are recognised as appropriate to the functioning of these supraspinal neurons. In this way physiological feedback is provided to the supraspinal neurons that alters their phenotype, allowing them to change from a growing to transmitting neurone, and protect them from cell death.

Even if no anatomical regeneration of nerve fibres is possible, it is still feasible that neuronal circuitry intrinsic to the spinal cord can provide some recovery of behavior in the absence of true regeneration as is seen, for instance, after spinal transection of the kitten, rat pup or frog. It seems that in these cases the locomotor generator neurons of the spinal cord can be stimulated into activity by sensory stimulation more easily than in the adult animal or the primate. However, in cats it is possible to stimulate these pattern generators pharmacologically.[154] Transplants of embryonic neurons below the transection site, which could substitute the missing neurotransmitters these pharmacological agents seek to mimic, has been attempted by several research groups and their results are described in more detail in separate chapter.

References

1. Bernstein JJ. Successful spinal cord regeneration: Known biological strategies. In: Reier PJ, Bunge RP, Seil FJ eds. Current Issues in Neural Regeneration Research. Vol 48. New York: Alan R Liss inc, 1988:331-341.
2. Stehouwer D. Behavior of larval and juvenile bullfrogs (Rana Catesbeiana) following chronic spinal transection. Behav Neural Biol 1986; 45:120-134.
3. Cummings J, Bernstein DR, Stelzner DJ. Further evidence for the sparing of function after spinal cord transection in the neonatal rat is not due to axonal generation or regeneration. Exp Neurol 1981; 74:615-620.
4. Stelzner DJ, Ershler WB, Weber ED. Effects of spinal transection in neonatal and weanling rats: Survival of function. Exp Neurol 1975; 46:156-177.
5. Weber ED, Stelzner DJ. Behavioral effects of spinal cord transection in the developing rat. Brain Res 1977; 125:241-255.
6. Lieberman AR. Some factors affecting retrograde neuronal responses to axonal lesions. In: Bellairs R, Gray EG, eds. Essays on the Nervous System. Oxford University Press, 1974:71-105.
7. Rovainen CM. Regeneration of Müller and Mauthner axons after spinal transection in larval lampreys. J Comp Neurol 1976; 168:545-554.
8. Selzer ME. Mechanisms of functional recovery and regeneration after spinal cord transection in the larval sea lamprey. J Physiol 1978; 277:395-408.
9. Wood MR, Cohen MJ. Synaptic regeneration in identified neurons of the lamprey spinal cord. Science 1979; 206:344-347.
10. Yin HS, Selzer ME. Axonal regeneration in lamprey spinal cord. J Neurosci 1983; 3:1135-1144.
11. Davis Jr GR, McClellan AD. Long Distance axonal regeneration of identified lamprey reticulospinal neurons. Exp Neurol 1994; 127:94-105.
12. Wood MR, Cohen MJ. Synaptic regeneration and glial reactions in the transected spinal cord of the lamprey. J Neurocytol 1981; 10:57-79.
13. McClellan AD. Functional regeneration of descending brainstem command pathways for locomotion demonstrated in the in vitro lamprey CNS. Brain Res 1988; 448:339-345.
14. McClellan AD. Brainstem command systems for locomotion in the lamprey: Localization of descending pathways in the spinal cord. Brain Res 1988; 457:338-349.
15. McClellan AD. Functional regeneration and recovery of locomotor activity in spinally transected lamprey. J Exp Zool 1992; 261:274-287.
16. Zhang L McClellan AD. Axonal regeneration of descending brain neurons in larval lamprey demonstrated by retrograde double labelling J Comp Neurol 1999; 410:612-626.
17. Zhang L, Palmer R, McClellan AD. Increase in descending brain-spinal cord projections with age in larval lamprey: Implication for spinal cord injury. J Comp Neurol 2002; 447:128-137.
18. Mackler SA, Selzer ME. Regeneration of functional synapses between individual recognizable neurons in the lamprey spinal cord. Science 1985; 229:774-776.
19. Mackler SA, Selzer ME. Specificity of synaptic regeneration in the spinal cord of the larval sea lamprey. J Physiol 1987; 388:183-198.
20. Cohen AH, Mackler SA, Selzer ME. Functional regeneration following spinal transection demonstrated in the isolated spinal cord of the larval sea lamprey. Proc Nat Acad Sci USA 1986; 83:2763-2766.

21. Cohen AH, Wallen P. The neuronal correlate of locomotion in fish "fictive swimming" induced in an in vitro preparation of the lamprey spinal cord. Exp Brain Res 1980; 41:11-18.

22. Cohen AH, Mackler SA, Selzer ME. Behavioral recovery following spinal transection: Functional regeneration in the lamprey CNS. Trends Neurosci 1988; 11:227-231.

23. Davis Jr GR, McClellan AD. Time course of anatomical regeneration of descending brainstem neurons and behavioral recovery in spinal-transected lamprey. Brain Res 1993; 602:131-137.

24. Lurie DI, Selzer ME. Axonal regeneration in the adult lamprey spinal cord. J Comp Neurol 1991; 306:409-416.

25. Bernstein JJ. Relation of spinal cord regeneration to age in adult goldfish. Brain Res 1964; 9:161-164.

26. Bernstein JJ, Bernstein ME. Effects of glial-ependymal scar and teflon arrest on the regenerative capacity of goldfish spinal cord. Exp Neurol 1967; 19:25-32.

27. Reier PJ, Stensaas LJ, Guth L. The astrocytic scar as an impediment to regeneration in the central nervous system. In: Kao CC, Bunge RP, Reier PJ, eds. Spinal Cord Reconstruction. New York: Raven Press, 1983:163-168.

28. Bernstein JJ, Geldered JB. Regeneration of long spinal tracts in the goldfish. Brain Res 1970; 20:33-38.

29. Bernstein JJ, Geldered JB. Synaptic reorganisation following regeneration of goldfish spinal cord. Exp Neurol 1973; 41:402-410.

30. Coggeshall RE, Youngblood CS. Recovery from spinal transection in fish: Regrowth of axons past the transection. Neurosci Letts 1983; 227-231.

31. Sharma SC, Jadhao PD, Rao PD. Regeneration of supraspinal projections neurons in the adult goldfish. Brain Res 1993; 620:221-228.

32. Hanna GF, Nawar NN, Sharma SC. Regeneration of ascending spinal axons in goldfish. Brain Res 1998; 791:235-245.

33. Becker T, Becker CG. Regenerating descending axons preferentially reroute to the gray matter in the presence of a general macrophage/microglia reaction caudal to a spinal transection in adult zebrafish. J Comp Neurol 2001; 433:131-147.

34. Becker T, Wullimann MF, Becker CG et al. Axonal regrowth after spinal cord transection in adult zebrafish. J Comp Neurol 1997; 377:577-595.

35. Becker T, Bernhardt RR, Reinhard E et al. Readiness of zebrafish brain neurons to regenerate a spinal axon correlates with differential expression of specific cell recognition molecules. J Neurosci 1998; 18:5789-5803.

36. McClellan AD. Spinal cord injury: Lessons from locomotor recovery and axonal regeneration in lower vertebrates. Neuroscientist 1998; 4:250-263.

37. Bernhardt RR. Cellular and molecular bases of axonal regeneration in the fish central nervous system. Exp Neurol 1999; 157:223-240.

38. Davis BM, Duffy MT, Simpson Jr SB. Bulbospinal and intraspinal connections in normal and regenerated salamander spinal cord. Exp Neurol 1989; 103:41-51.

39. Benraiss A, Arsanto JP, Coulon J et al. Neurogenesis during caudal spinal cord regeneration in adult newts. Dev Genes Evol 1999; 209:363-369.

40. Forehand CJ, Farel PB. Anatomical and behavioral recovery from the effects of spinal cord transection: Dependence on metamorphosis in anuran larvae. J Neurosci 1982; 2:654-662.

41. Beattie MS, Bresnahan JC, Copate G. Metamorphosis alters the response to spinal cord transection in Xenopus laevis frogs. J Neurobiol 1990; 21:1108-1122.

42. Lee MT. Regeneration and functional reconnection of an identified vertebrate central neuron. J Neurosci 1982; 2:1793-1811.

43. Brenner PR, Stehouwer DJ. Sparing and recovery of function in spinal larval frogs (Rana Catesbeiana): Effect of level of transection. Behav Neurol Biol 1991; 56:292-306.

44. Ramon y Cajal S. (translated by RM May) Degeneration and regeneration of the nervous system. Oxford:Oxford University press, 1928.

45. Fry EJ, Saunders NR. Spinal repair in immature animals: A novel approach using the South American opossum Monodelphis domestica. Clin Exp Pharmacol Physiol 2000; 27:542-547.

46. Fawcett JW. Intrinsic neuronal determinants of regeneration. Trends Neurosci 1992; 15:5-8.

47. Terman JR, Wang XM, Martin GF. Repair of the transected spinal cord at different stages of development in the North American opossum, Didelphis virginiana. Brain Res Bull 2000; 53:845-855.

48. Varga ZM, Bandtlow CE, Erulkar SD et al. The critical period for repair of CNS of neonatal opossum (Monodelphis domestica) in culture: Correlation with development of glial cells, myelin and growth-inhibitory molecules. Eur J Neurosci 1995; 7:2119-2129.

49. Saunders NR, Kitchener P, Knott GW et al. Development of walking, swimming and neural connections after completed spinal cord transection in the neonatal opossum, Monodelphis domestica. J Neurosci 1998; 18:339-355.

50. Saunders NR, Adams E, Reader M et al. Monodelphis domestica (grey short-tailed opossum): An accessible model for studies of early neocortical development. Anat Embryol (Berl) 1989; 180:227-236.

51. Treherne JM, Woodward SK, Varga ZM et al. Restoration of conduction and growth of axons through injured spinal cord of neonatal opossum in culture. Proc Natl Acad Sci USA 1992; 89:431-434.

52. Nicholls J, Saunders N. Regeneration of immature mammalian spinal cord after injury. Trends Neurosci 1996; 19:229-234.

53. Woodward SK, Treherne JM, Knott GW et al. Development of connections by axons growing through injured spinal cord of neonatal opossum in culture. J Exp Biol 1993; 176:77-88.

54. Varga ZM, Schwab ME, Nicholls JG. Myelin-associated neurite growth-inhibitory proteins and suppression of regeneration of immature mammalian spinal cord in culture. Proc Natl Acad Sci USA 1995; 92:10959-10963.

55. Saunders NR, Deal A, Knott GW et al. Repair and recovery following spinal cord injury in a neonatal marsupial (Monodelphis domestica). Clin Exp Pharmacol Physiol 1995; 22:518-526.

56. Varga ZM, Fernandez J, Blackshaw S et al. Neurite outgrowth through lesions of neonatal opossum spinal cord in culture. J Comp Neurol 1996; 366:600-612.

57. Wang XM, Xu XM, Qin YQ et al. The origins of supraspinal projections to the cervical and lumbar spinal cord at different stages of development in the gray short-tailed Brazilian opossum, Monodelphis domestica. Dev Brain Res 1992; 68:203-216.

58. Holst M, Ho RH, Martin GF. The origins of supraspinal projections to lumbosacral and cervical levels of the spinal cord in the gray short-tailed Brazilian opossum, Monodelphis domestica. Brain Behav Evol 1991; 38:273-289.

59. Weber ED, Stelzner DJ. Synaptogenesis in the intermediate gray region in the lumbar spinal cord in the postnatal rat. Brain Res 1980; 185:17-37.

60. Navarrete R, Vrbová G. Activity-dependent interactions between motoneurons and muscles: Their role in the development of the motor unit. Prog Neurobiol 1993; 41:93-124.

61. Donatelle JM. Growth of the corticospinal tract and the development of placing reactions in the postnatal rat. J Comp Neurol 1977; 175:207-231.

62. Gribnau AA, de Kort EJ, Dederen PJ et al. On the development of the pyramidal tract in the rat II. An anterograde tracer study of the outgrowth of the corticospinal fibres. Anat Embryol 1986; 175:101-110.

63. Bregman BS. Development of serotonin immunoreactivity in the rat spinal cord and its plasticity after neonatal spinal cord lesions. Dev Brain Res 1987; 34:245-263.

64. Chen KS, Stanfield BB. Evidence that selective collateral elimination during postnatal development results in a restriction in the distribution of locus coeruleus neurons which project to the spinal cord in rats. Brain Res 1987; 410:154-158.

65. Gilbert M, Stelzner DJ. The development of descending and dorsal root connections in the lumbosacral spinal cord of the postnatal rat. J Comp Neurol 1979; 184:821-838.

66. Hulsebosch CE, Coggeshall RE. A comparison of axonal numbers in dorsal root following spinal cord hemisection. Brain Res 1983; 265:187-197.

67. Robinson GA, Goldberger ME. The development and recovery of motor function in spinal cats. I. The infant lesion effect. Exp Brain Res 1986; 62:373-386.

68. Bradley NS, Smith KL. Neuromuscular patterns of stereotypic hindlimb behaviors in the first two post-natal months. II Stepping in spinal kittens. Dev Brain Res 1988; 38:53-67.

69. Goldberger ME. The use of behavioral methods to predict spinal cord plasticity. Restor Neurol Neurosci 1991; 2:339-350.

70. Smith JL, Smith LA, Zernicke RF et al. Locomotion in exercised and nonexercised cats cordotomised at 2 and 12 weeks. Exp Neurol 1982; 76:393-413.

71. Goldberger ME, Murray M. Recovery of function and anatomical plasticity after damage to adult and neonatal spinal cord. In: Cotman CW, ed. Synaptic Plasticity. New York: Guilford, 1985:77-110.

72. Robinson GA, Goldberger ME. The development and recovery of motor function in spinal cats. II. Pharmacological enhancement of recovery. Exp Brain Res 1986; 62:387-400.

73. Bernard JW, Carpenter W. Lack of regeneration in spinal cord of rat. J Neurophysiol 1950; 13:223-228.

74. Björklund A, Katzman R, Stenevi U et al. Development and growth of axonal sprouts from noradrenaline and 5-hydroxytryptamine neurons in the rat spinal cord. Brain Res 1971; 31:21-33.

75. Prendergast J, Shusterman R. Normal development of motor behavior in the rat and the effect of midthoracic spinal hemisection at birth on that development. Exp Neurol 1982; 78:176-189.

76. Skagerberg G, Björklund A. Topographic principles in the spinal projections of serotonergic and nonserotonergic brainstem neurons in the rat. Neuroscience 1985; 15:445-480.

77. Martin GF, Xu XM. Evidence for the developmental plasticity of the rubrospinal tract. Studies using the North American opossum. Dev Brain Res 1988; 39:303-308.

78. Kalil K, Reh T. Regrowth of severed axons in the neonatal central nervous system: Establishment of normal connections. Science 1979; 205:1158-1161.

79. Kuang RZ, Kalil K. Branching patterns of corticospinal axon arbors in the rodent. J Comp Neurol 1990; 292:585-598.

80. Liu CN, Chambers WW. Intraspinal sprouting of dorsal root axons. Arch Neurol Psychiat 1958; 79:46-61.

81. Bullitt E, Stofer WD, Vierek CJ et al. Reorganization of primary afferent nerve terminals in the spinal dorsal horn of the primate caudal to anterolateral chordotomy. J Comp Neurol 1988; 270:549-558.

82. Polistina DC, Murray M, Goldberger ME. Plasticity of dorsal root and descending serotonergic projections after partial deafferentation of the rat spinal cord. J Comp Neurol 1990; 299:349-363.

83. Rodin BE, Sampogna SL, Kruger L. An examination of intraspinal sprouting in dorsal root axons with tracer HRP. J Comp Neurol 1983; 215:187-198.

84. Pubols LM, Bowen DC. Lack of central sprouting of primary afferent fibres after ricin deafferentation. J Comp Neurol 1988; 275:282-287.

85. McMahon SB, Kett-White R. Sprouting of peripherally regenerating primary sensory neurons in the adult central nervous system. J Comp Neurol 1991; 304:307-315.

86. Fitzgerald M. The sprouting of saphenous nerve terminals in the spinal cord following early postnatal sciatic nerve section in the rat. J Comp Neurol 1985; 240:407-413.

87. Fitzgerald M, Vrbová G. Plasticity of acid phosphatase (FRAP) afferent terminal fields and of dorsal horn cell growth in the neonatal rat. J Comp Neurol 1985; 240:414-420.

88. Réthelyi M, Salim MZ, Jancso G. Altered distribution of dorsal root fibres in the rat following neonatal capsaicin treatment. Neuroscience 1988; 18:749-761.

89. Fitzgerald M, Woolf CJ, Shortland P. Collateral sprouting of the central terminals of cutaneous primary afferent neurons in the rat spinal cord: Pattern, morphology, and influence of targets. J Comp Neurol 1990; 300:370-385.

90. Smith CL. The development and postnatal organization of primary afferent projections to the rat thoracic spinal cord. J Comp Neurol 1983; 220:29-43.

91. Fitzgerald M, Swett J. The termination pattern of sciatic nerve afferents in the substantia gelatinosa of neonatal rats. Neurosci Lett 1983; 43:149-154.

92. Fitzgerald M. The post-natal development of cutaneous afferent fibre input and receptive field organization in the rat dorsal horn. J Physiol 1985; 364:1-18.

93. Richardson PM, Issa VM. Peripheral nerve injury enhances central regeneration of primary sensory neurons. Nature 1984; 309:791-793.

94. Woolf CJ, Shortland P, Coggeshall RE. Peripheral nerve injury triggers central sprouting of myelinated afferents. Nature 1992; 355:75-78.

95. Florence SL, Garraghty PE, Carlson M et al. Sprouting of peripheral nerve axons in the spinal cord of monkeys. Brain Res 1993; 601:343-348.

96. Skene JHP. Axonal growth-associated proteins. Annu Rev Neurosci 1989; 12:127-156.

97. Woolf CJ, Reynolds ML, Molander C et al. The growth-associated protein GAP43 appears in dorsal root ganglion cells and in the dorsal horn of the rat spinal cord following peripheral nerve injury. Neuroscience 1990; 34:465-478.

98. Helgren ME, Goldberger ME. The recovery of postural reflexes and locomotion following low thoracic hemisection in adult cats involves compensation by undamaged primary afferent pathways. Exp Neurol 1993; 123:17-34.

99. Goldberger ME. Locomotor recovery after unilateral hindlimb deafferentation in cats. Brain Res 1977; 123:59-74.

100. Goldberger ME. Spared root deafferentation of a cat's hindlimb: Hierarchical regulation of pathways mediating recovery of motor behavior. Exp Brain Res 1988; 73:329-342.

101. Murray M, Goldberger ME. Replacement of synaptic terminals in lamina II and Clarke's nucleus after unilateral lumbosacral rhizotomy in adult cats. J Neurosci 1986; 6:3205-3217.

102. Zhang B, Goldberger ME, Murray M. Proliferation of SP and 5HT-containing terminals in lamina II of rat spinal cord following dorsal rhizotomy: Quantitative EM-immunocytochemical studies. Exp Neurol 1993; 123:51-63.

103. Tessler A, Glazer E, Artymyshyn R et al. Recovery of substance P in the cat spinal cord after unilateral lumbosacral deafferentation. Brain Res 1980; 191:459-470.

104. Tessler A, Himes BT, Artymyshyn R et al. Spinal neurons mediate return of substance P following deafferentation of cat spinal cord. Brain Res 1981; 230:263-281.

105. Beattie MS, Leedy MG, Bresnahan JC. Evidence for alterations of synaptic inputs to sacral spinal reflex circuits after spinal cord transection in the cat. Exp Neurol 1993; 123:35-50.

106. Lahr SP, Stelzner DJ. Anatomical studies of dorsal column axons and dorsal root ganglion cells after spinal cord injury in newborn rat. J Comp Neurol 1990; 293:377-398.

107. Bates CA, Killackey HP. The emergence of a discretely distributed pattern of corticospinal projection neurons. Dev Brain Res 1984; 13:265-273.

108. Himes BT, Tessler A. Death of some dorsal root ganglion neurons and plasticity of others following sciatic nerve section in adult and neonatal rats. J Comp Neurol 1989; 284:215-230.

109. Bregman BS, Bernstein-Goral H, Kunkel-Bagden E. CNS transplants promote anatomical plasticity and recovery of function after spinal cord injury. Restor Neurol Neurosci 1991; 2:327-338.

110. Yip HK, Johnson Jr EM. Developing dorsal root ganglion neurons require trophic support from their central processes: Evidence for a role of retrogradely transported nerve growth factor from the central nervous system to the periphery. Proc Natl Acad Sci USA 1984; 81:6245-6249.

111. Perkins S, Carlstedt T, Mizuro K et al. Failure of regenerating dorsal root axons to regrow into the spinal cord. Can J Neurol Sci 1980; 7:323-332.

112. Anderson PN, Chong MS, Woolf CM et al. GAP43 and lumbar dorsal root regeneration into long peripheral nerve grafts in rat. Neurosci Lett Suppl 1992; 42:S12.

113. Carlstedt T. Reinnervation of the mammalian spinal cord after neonatal dorsal root crush. J Neurocytol 1988; 17:335-350.

114. Sah DW, Frank E. Regeneration of sensory-motor synapses in the spinal cord of the bullfrog. J Neurosci 1984; 4:2784-2791.

115. Frank E, Sah DW. Reformation of specific synaptic connections by regenerating sensory axons in the spinal cord of the bullfrog. Neurochem Pathol 1986; 5:165-185.

116. Kliot M, Smith GM, Siegal JD et al. Astrocyte-polymer implants promote regeneration of dorsal root fibers into adult mammalian spinal cord. Exp Neurol 1990; 109:57-69.

117. Prendergast J, Stelzner DJ. Changes in the magnocellular portion of the red nucleus following thoracic hemisection in the neonatal and adult rat. J Comp Neurol 1976; 166:163-172.

118. Bregman BS, Reier PJ. Neural tissue transplants rescue axotomised rubrospinal cells from retrograde death. J Comp Neurol 1986; 244:86-95.

119. McBride RL, Feringa ER, Garver MK et al. Relabelled red nucleus and sensorimotor cortex neurons of the rat survive 10 and 20 weeks after spinal cord transection. J Neuropathol Exp Neurol 1989; 48:568-576.

120. McBride RL, Feringa ER, Garver MK et al. Retrograde transport of fluoro-gold in corticospinal and rubrospinal neurons 10 and 20 weeks after T9 spinal cord transection. Exp Neurol 1990; 108:83-85.

121. Lowrie MB, Vrbová G. Dependence of post-natal motoneurons on their targets: Review and hypothesis. Trends Neurosci 1992; 15:80-84.
122. Merline M, Kalil K. Cell death of corticospinal neurons is induced by axotomy before but not after innervation of spinal targets. J Comp Neurol 1990; 296:506-516.
123. Bregman BS, Bernstein-Goral H. Both regenerating and late developing pathways contribute to transplant-induced anatomical plasticity after spinal cord lesions at birth. Exp Neurol 1991; 112:49-63.
124. Schreyer DT, Jones EG. Growth and target finding by axons of the corticospinal tract in pre natal and postnatal rats. Neuroscience 1982; 7:1837-1853.
125. Tolbert DL, Der T. Redirected growth of pyramidal tract axons following neonatal pyramidotomy in cats. J Comp Neurol 1987; 260:299-311.
126. Bregman BS, Goldberger ME. Infant lesion effect: III Anatomical correlates of sparing and recovery of function after spinal cord damage in newborn and adult cats. Dev Brain Res 1983; 9:137-154.
127. Bates CA, Stelzner DJ. Extension and regeneration of corticospinal axons after early spinal injury and the maintenance of corticospinal topography. Exp Neurol 1993; 123:106-117.
128. Bregman BS, Kunkel-Bagden E, McAtee M et al. Extension of the critical period for developmental plasticity of the corticospinal pathway. J Comp Neurol 1989; 282:355-370.
129. Joosten EA, Gribnau AA, Dederen PJ. An anterograde study of the developing corticospinal tract in the rat: Three components. Dev Brain Res 1987; 36:121-130.
130. Kunkel-Bagden E, Bregman BS. Spinal cord transplants enhance the recovery of locomotor function after spinal cord injury at birth. Exp Brain Res 1990; 81:25-34.
131. Diener PS, Bregman BS. Fetal spinal cord transplants support the development of target reaching and coordinated postural adjustment after neonatal cervical cord injury. J Neurosci 1998; 18:763-778.
132. Diener PS, Bregman BS. Fetal spinal cord transplants support growth of supraspinal and segmental projections after cervical spinal cord hemisaction in the neonatal rat. J Neurosci 1998; 18:779-793.
133. Hase T, Kawaguchi S, Hayashi H et al. Spinal cord repair in neonatal rats: A correlation between axonal regeneration and functional recovery. Eur J Neurosci 2002; 15:969-974.
134. Rosenthal BM, Alley KE. Trigeminal motoneurons in frogs develop a new dendritic field during metamorphosis. Neurosci Lett 1988; 95:53-58.
135. Bray GM, Villegas-Perez MP, Vidal-Sanz M et al. The use of peripheral nerve grafts to enhance neuronal survival, promote growth and permit terminal connections in the central nervous system of adult rats. J Exp Biol 1987; 132:5-19.
136. Lurie DI, Selzer ME. Preferential regeneration of spinal axons through the scar in hemisected lamprey spinal cord. J Comp Neurol 1991; 313:669-679.
137. Mansour H, Asher R, Dahl D et al. Permissive and nonpermissive reactive astrocytes: Immunofluorescence study with antibodies to the glial hyaluronate-binding protein. J Neurosci Res 1990; 25:300-311.
138. Bovolenta P, Wandosell F, Nieto-Sampedro M. Neurite outgrowth over resting and reactive astrocytes. Restor Neurol Neurosci 1991; 2:221-228.
139. Pixley SK, Nieto-Sampedro M, Cotman CW. Preferential adhesion of brain astrocytes to laminin and central neurites to astrocytes. J Neurosci Res 1987; 18:402-406.
140. Michel ME, Reier PJ. Axonal-ependymal associations during early regeneration of the transected spinal cord in xenopus laevis tadpoles. J Neurocytol 1979; 8:529-548.
141. Stensaas LJ. Regeneration in the spinal cord of the newt Noptopthalmus. In: Kao CC, Bunge RP, Reier PJ, eds. Spinal Cord Reconstruction. New York: Raven Press, 1983:121-150.
142. Bandtlow C, Zachleder T, Schwab ME. Oligodendrocytes arrest neurite growth by contact inhibition. J Neurosci 1990; 10:3837-3848.
143. Caroni P, Schwab ME. Two membrane protein functions from rat central myelin with inhibitory properties for neurite growth and fibroblast spreading. J Cell Biol 1988; 106:1281-1288.
144. Keirstead HS, Hasan SJ, Muir GD et al. Suppression of the onset of myelination extends the permissive period for the functional repair of embryonic spinal cord. Proc Natl Acad Sci USA 1992; 89:11664-11668.
145. Prendergast J, Misantone LJ. Sprouting of tracts descending from midbrain to spinal cord: The result of thoracic funicolotomy in newborn, 21d and adult rat. Exp Neurol 1980; 69:458-480.

146. Ang LC, Bhaumick B, Munoz DG et al. Effects of astrocytes, insulin and insulin-like growth factor 1 on the survival of motoneurons in vitro. J Neurol Sci 1992; 109:168-172.

147. Skaper SD, Varon S. Age-dependent control of dorsal root ganglion neuron survival by macromolecular and low-molecular weight trophic agents and sub-stratum bound laminins. Dev Brain Res 1986; 24:39-46.

148. Heumann R, Korsching S, Bandtlow C et al. Changes of nerve growth factor synthesis in nonneuronal cells in response to sciatic nerve transection. J Cell Biol 1987; 104:1623-1631.

149. Meyer M, Matsuoka I, Wetmore C et al. Enhanced synthesis of brain-derived neurotrophic factor in the lesioned peripheral nerve: Different mechanisms are responsible for the regulation of BDNF and NGF mRNA. J Cell Biol 1992; 119:45-54.

150. Taniuchi M, Clark HB, Schweitzer JB et al. Expression of nerve growth factor receptors by Schwann cells of axotomised peripheral nerves: Ultrastructural location, suppression by axon contact, and binding properties. J Neurosci 1988; 8:664-681.

151. Snider WD, Elliot JL, Yan Q. Axotomy-induced neuronal death during development. J Neurobiol 1992; 23:1231-1246.

152. Joosten EA, Van der Ven PF, Hooiveld MH et al. Induction of corticospinal target finding by release of a diffusable, chemotropic factor in cervical spinal grey matter. Neurosci Letts 1991; 128:25-28.

153. Greensmith L, Vrbová G. Alterations of nerve-muscle interaction during postnatal development influence motoneurone survival in rats. Dev Brain Res 1992; 69:125-131.

154. Barbeau H, Rossignol S. Initiation and modulation of the locomotor pattern in the adult chronic spinal cat by noradrenergic, serotonergic and dopaminergic drugs. Brain Res 1991; 546 250-260.

CHAPTER 3

Recovery of Lost Spinal Cord Function by Facilitating the Spinal Cord Circuits Below the Lesion

Urszula Slawinska

Introduction

I n the previous chapter strategies restore the lost function after spinal cord injury (a) by attempting to reconnect anatomically the separated parts of the spinal cord by encouraging regeneration of axons across the damaged parts of the spinal cord; (b) by providing grafts that would act as "relays"; (c) replacing specific populations of cells that may have been damaged during the injury were reviewed.

Another approach that would use the neural circuitry below the lesion and encourage its function has been little explored. The first attempt to enhance the activity of spinal cord circuitry below lesion employed the systemic administration of agonists of monoamines that modified locomotor performance of paraplegic cats.[1-3] The other strategy that could be used for inducing the enhancement of the activity of the neural spinal cord circuitry below the lesion is intraspinal grafting of embryonic neurons of supraspinal origin. Selected embryonic neurons can be grafted in order to replace missing supraspinal input after complete spinal cord transection. Among these supraspinal inputs, serotoninergic projection from raphe neurons and noradrenergic projection from locus coeruleus have been extensively studied during the last three decades. The present chapter will review findings that introduced the promising grafting strategy to enhance the restoration of some function (like hindlimb locomotor movement or sexual reflexes) which are dramatically altered after complete spinal cord transection in rats.

Spinal Cord Functions Controlled by Brain Stem Monoaminergic Connections

The neural structures of the spinal cord, that control the hindlimb locomotor movement or sexual reflexes, receive information from the periphery as well as from the descending projection from supraspinal brain structures. Among the descending supraspinal projections the monoaminergic systems play a crucial role in the control of the excitability of the spinal circuitry.

Transplantation of Neural Tissue into the Spinal Cord, Second Edition, edited by Antal Nógrádi. ©2006 Eurekah.com and Springer Science+Business Media.

Localization of Cell Bodies and Descending Pathways

The supraspinal monoamine-containing clusters of neurons are located near the middle of the pons and upper brainstem. The main source of noradrenergic fibres arises from the locus coeruleus, subcoeruleus and a few other regions of the pons.[4-6] Out of the five major noradrenergic tracts one descends into the mesencephalon and the spinal cord, where the fibres course in the ventral-lateral column. The other four (ascending tracts) innervate different parts of the cerebral cortex, specific thalamic and hypothalamic nuclei, the olfactory bulb and the cerebellar cortex. The serotonin-containing neurons are known to be restricted to clusters of cells lying in or near the middle or the raphe regions of the pons and upper brain stem. The serotonergic cell bodies that belong to the anterior raphe nuclei (nucleus raphe dorsalis, centralis superior, and linearis rostralis) send fibres to forebrain areas, i.e., striatum, hippocampus, hypothalamus, septum, amygdala and cerebral cortex. Whereas those distributed in the posterior raphe nuclei (nucleus raphe obscurus, pallidus, and magnus) send the projections invading the spinal cord.[7] Skagerberg and Björklund[5] described three fairly distinct pathways for spinal serotonergic projections from medullary raphe and adjacent regions: dorsal, intermediate and ventral pathways. The dorsal pathway originates mainly from cells in the caudal pons and rostral medulla oblongata. This pathway descends through the dorsal part of the lateral funiculus and terminates mainly in the dorsal horn at all spinal levels. The intermediate pathway with cell bodies located in the nucleus raphe obscurus and pallidus, as well as within the accurate cell group (situated just ventral and lateral to the pyramids very close to the ventral surface of the brainstem), terminates in the intermediate grey matter at thoracolumbar and upper sacral levels. The ventral pathway descends in the ventral white matter to terminate in ventral areas along the entire length of the spinal cord. These axons originate mainly from neurons in the caudal medulla. Thus, in the spinal cord the serotoninergic as well as noradrenergic fibres and terminals are concentrated in the dorsal horn (lamina I and outer part of lamina II), in the intermedialateral cell column at thoracic and sacral levels, and in the ventral horn near the different groups of motoneurons.[5,7-9]

Functional Implications

Spinal serotoninergic projections have been described in 1960 by Brodal et al.[10] However, at this time the functional implications of these projections were obscure. Since then many authors speculated on the involvement of descending serotonergic system in somatic and autonomic motor function and also in influencing reflex responses to sensory stimuli.[11,12] Further investigations confirmed that descending serotonergic as well as noradrenergic pathways make synaptic contact e.g., with motoneurons.[13,14] Moreover, electrophysiological studies established that monoamines in the spinal cord modulate synaptic gain and adjust the level of spinal cord excitability. There is considerable evidence to suggest that serotonin or noradrenaline release from descending nerve terminals of these fibres facilitates spinal motoneurone excitability.[15-21] Moreover, these monoamines applied locally modulate the transmission between primary afferents and premotor interneurons. This has a particularly strong impact on reflex actions because these interneurons directly excite or inhibit motoneurons.[22] In this way the supraspinal structures can modulate particular function controlled by the spinal cord neural circuitry.[23]

An example of a particular function controlled by the neural circuitry of the spinal cord is locomotion. It was established some time ago that the locomotor functions in the cat,[18,19,24-27] rat[28-30] as well as in monkey[31] and human[32,33] are controlled by the neural circuitry of the spinal cord. There is evidence that locomotor movement of hindlimbs might be elicited even if the spinal cord neural circuitry is separated from supraspinal inputs. However, the descending monoaminergic supraspinal projections are very important for the regulation of the spinal reflex activities and the modulation of the spinal locomotor pattern generating mechanisms (for review see ref. 34). Early studies showed that acute spinal cats could generate fictive locomotion after intravenous injection of L-Dopa[18,19,25] and stepping movement on a treadmill after administration of the α_2-noradrenergic agonist clonidine.[35] It is evident that the monoaminergic projections regulate the lumbosacral neuronal circuitry (including interneurons and motoneurons) that control hindlimb movements.[20,21,23] At least two descending monoaminergic systems are thought to provide tonic stimulation of the spinal neural circuitry that controls locomotion. As mentioned above, one originates in the locus coeruleus and releases noradrenaline (NA),[36] the other arises predominately in the raphe nuclei of the medulla oblongata and releases serotonin (5-hydroxytryptamine, 5-HT).

Urogenital functions are also controlled by neural circuitry of the spinal cord. In the rats, the spinal cord at the L6-S1 level contains the autonomic preganglionic neurons that control urogenital function, e.g., penile erection. Moreover, the ventral horn of the L5-L6 segment of the spinal cord contains motoneurons that supply pelvic striated muscles, e.g., the ischiocavernosus and bulbospongiosus muscles.[37,38] These particular neurons receive strong serotonergic projections that control penile erection. Stimulation of spinal serotonergic receptors suppresses reflex erections and facilitates ejection of semen in intact rats.[39] It was shown that spinal cord transection, electrolytic lesion of the nucleus paragigantocellularis (the main source of descending 5-HT projections to the spinal cord) and chemical lesion of 5-HT pathways altered the reflex erection in rats.[40-43] The crucial role of 5-HT in sexual reflexes has also been reinforced by pharmacological experiments in which the penile erection was elicited by m-chlorophenylpiperazine, a $5\text{-HT}_{2B/2C}$ and 5-HT_{1B} receptor agonist.[44,45]

The reduced release of noradrenaline and serotonin after spinal cord injury alters markedly the locomotor functions in mammals. Total spinal cord transection induces almost complete disappearance of monoaminergic neurotransmiters below the lesion. Two weeks after a severe injury to the rat spinal cord, there is almost complete loss of 5-HT in the lumbar spinal cord, and the animals displayed severe spastic paraparesis.[46] In these experiments, reduction in immunocytochemical staining of serotonin in the ventral grey matter of the lumbar region of the spinal cord at 2 weeks after trauma was correlated with the degree of the severity of the insult as reflected by motor impairment. The motor scores correlated significantly with changes in serotonin staining in the ventral but not in the dorsal horn.[46] Moreover, the experiments involving microdialysis documented that serotonin release is significantly lower 4 weeks after incomplete spinal cord injury.[47] It means that after complete as well as incomplete spinal cord injuries, the descending 5-HT fibres in the lateral and anterior funiculi undergo a slow wallerian degeneration. Moreover, it was also demonstrated that the recovery of motor function after spinal cord injury is correlated with the level of monoamines remaining in the spinal cord.[48,49] The systemic administration (intravenously or intraperitoneally) of agonists of monoamines modifies spinal reflexes and improves locomotor performance of animals which had their spinal cord transected.[1-3] After spinal cord transection, extensor excitability is sharply reduced and the stretch reflex is virtually abolished, but can be restored by subsequent administration of serotonin precursors or agonists.[12,50] In humans with spinal cord injury (SCI), clonidine, a noradrenergic agonist, and cyproheptadine, a serotonergic antagonist, have each been associated with improvement walking in SCI subjects.[51] Moreover, after such treatment most subjects with SCI have shown a reduction in signs and symptoms of spasticity.[52-54]

It is known that the bladder function also requires supraspinal control. During normal micturation there is a coordinated activation of smooth muscle of the bladder (detrusor) and striated muscle of the external urethral sphincter. This is achieved by the integration of excitatory, inhibitory and sensory nerve activity in control centres of the spinal cord, pons and forebrain. The bladder contraction may occur through the stimulation of locus coeruleus in the micturition centre and consequent release of noradenaline in the spinal cord.[55] On the other hand, electrical stimulation of 5-HT containing neurons in the caudal raphe and activation of postsynaptic 5-HT receptors in the spinal cord, via the release of 5-HT, inhibits bladder contractions.[56] Following spinal cord injury the urinary tract functions are dramatically altered. Bladder contractions produce high intravesical pressures but are not sustained and do not produce bladder emptying. It was shown that exogenous 5-HT was able to ameliorate bladder hyperactivity in spinally transected cats. Thus, compounds that would reduce bladder activity would be useful in protecting the kidneys against pressureinduced damage.

However, the systemic administration of monoaminergic drugs not only positively modulates the central nervous system functions, but also alters the function of the peripheral autonomic nervous system and this has serious side effects.[57] For example oral or intravenous clonidine application causes bradycardia and hypotension, which limits its clinical use for treating spasticity. Therefore various methods for localised intraspinal application of monoaminergic drugs were recently carefully explored. Such localized release could be accomplished by grafts of cells that release monoaminergic neurotransmitters. Embryonic cells taken from the brain stem grafted locally into the spinal cord can survive and release monoamines. It can therefore be assumed that neurotransmiters, such as serotonin or noradrenaline, released from intraspinally grafted monoaminergic neurons would be able to stimulate locally neural structures and in this way regulate their functions. It might be that the complex sequences of the locomotor-like hindlimb movement as well as the penile reflexes that are both dramatically altered after spinal cord injury in mammals could be significantly restored using this therapy.

Intraspinal Transplantation of Neural Cells of Supraspinal Origin

Methods Used

The early findings of development of monoaminergic reinnervation of the transected spinal cord by homologous fetal brain grafts was published almost 25 years ago.[58] It was shown that immature noradrenaline (NA) containing cells from locus coeruleus and 5-hydroxytryptamine (5-HT) containing cells from raphe nuclei could survive homologous transplantation to adult spinal cord that were adrenergically and serotonergically denervated by a transverse lesion. Moreover, both the NA and the 5-HT transplants contained the fluorescent cell bodies from which both axons in the white matter and nerve terminals in the grey matter had grown cranially and caudally in the spinal cord. The NA and 5-HT containing nerve terminals in the grey matter had normal-looking morphology with varicosities and intervaricose parts similar to those seen in intact spinal cords. Moreover, the new fibres were seen to innervate motoneurons and their dendrites in a normal way.[58]

However the most important question is whether the grafted cells can restore the lost function due to the spinal cord injury. Initially most researchers focussed their attention on the morphological study of the features of the grafted cells and their connections. Two different methods of transplantation were developed to investigate the survival of the grafted cells: (1) a solid piece of embryonic tissue was grafted into the lumbar spinal cord;[48,58,59] (2) isolated cell suspensions were used for intraspinal grafting.[59-64] In both cases, the defined region of the brain stem containing the raphe and/or locus coeruleus nuclei was dissected from embryonic neural tissue. Before transplantation, the spinal cord of all the host animals was deprived of their natural monoaminergic supraspinal innervation. It was believed that the destruction of natural monoaminergic innervation might provide better condition for survival of grafted embryonic tissue containing monoaminergic cells in the host environment. To destroy monoaminergic innervation of the spinal cord either a complete spinal cord transection was carried out or monoaminergic cells were selectively destroyed by administration of 6-hydroxydopamine (6-OHDA) or 5,7-dihydroxytryptamine (5,7 DHT). These compounds are known to destroy selectively monoaminergic cells.[60,62,65]

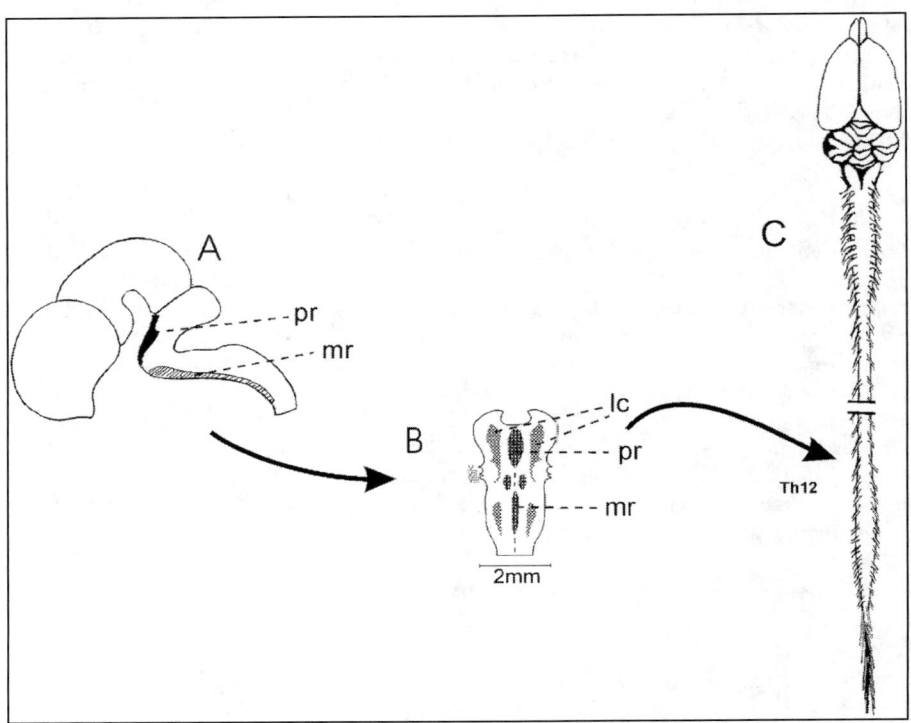

Figure 1. Diagram presents schematically the consecutive steps of preparation and intraspinal transplantation of the embryonic neural tissue. A) The embryonic forebrain with the brainstem. B) The dissected brainstem. Note the various regions of brainstem that are dissected for transplantation by different authors mention in this chapter. The source of serotonergic cells is pontine raphe nucleus (pr) or medullary raphe nucleus (mr). The noradrenergic cells are dissected from locus coeruleus (lc). C) the place of intraspinal trasnplantation used by different authors is mainly located in middle segments of thoracic spinal cord (Th10-Th12).

For intraspinal grafting of rats the embryonic donor tissue was dissected from rat pups (E14-E15) of the same inbred strain as the host (for methods see refs. 64,66-68). This age of fetuses corresponds to the time when most of the neurogenesis in the embryonic brain stem nuclei is completed. After the midbrain exposure, either the brain stem between the pontine and mesencephalic flexures (in order to isolate the locus coeruleus or rostral rhombencephalic raphe region) or the brain stem extending from pontine flexure to the cervical end of the spinal cord (to isolate the caudal rhombencehalic raphe region) were prepared (Fig. 1). A solid piece of tissue was used for grafting as a rectangular block of 0.5 mm containing the selected parts of brainstem.[58,68-70] To prepare the cell suspension for grafting, similar pieces of dissected tissue taken from the brain stem were mechanically dissociated.[60,71-74] The suspension prepared from nucleus coeruleus contained about 30 000 cell/µl out of which 1 to 2% were noradrenergic.[75]

The size and location of the graft in the spinal cord varied in experiments done by different authors. Also the time of grafting after destroying the monoaminergic innervation of the host spinal cord differed in various experiments. Mostly grafting procedures were carried out within 1 week after spinal cord injury, but recent results show that even when the tissue was grafted one month after spinal cord transection it survived and made functional connections with the host neural circuitry.[68] Thus, the graft is able to survive and develop equally well one week or one month after spinal cord injury.

Whether the two different techniques, i.e., solid grafts or cell suspensions, provided comparable good survival of grafted cells as well as the development of the monoaminergic reinnervation of the host spinal cord will be presented in the next section.

Survival of Grafted Cells

Survival of the monoaminergic neurons and the presence of their axons in the spinal cord were verified using several immunocytochemical and histofluorescent methods. It was shown that in the case of the locus coeruleus about 90% of the solid subpial grafts survived and contained up to 800-1100 surviving noradrenergic neurons. Since two pieces of locus coeruleus were grafted and one locus coeruleus contains about 1500 neurons, about 30% of the grafted neurons survived.[59,60,76,77] The noradrenergic fibres that grew from the graft extended rostrally and caudally, primarily within the grey matter, along the spinal cord for distance of up to 11 mm.[59]

It is important to note that in the spinal cord of adult mammals the only source of monoaminergic fibres are axons of neurons from supraspinal structures. Thus, the spinal cord of rats below the site of a complete lesion is totally devoid of monoaminergic innervation. However, following intraspinal transplantation of the raphe nuclei the 5-HT- immunoreactive neurons were visible in the vicinity of the graft. The morphology of these neurons was very similar to that in the dorsal or median raphe. All those neurons were of the stellate type, with up to four first order dendrites spreading from the perikaryon. Most of the dendrites were relatively straight, although in some cases they were highly branched. The size of the serotonergic neurons in the raphe and transplant was similar (their mean longest diameter in the raphe nuclei was 14.4 $\pm$1.7 μm ($\pm$ SD), and in the transplant was 14.1$\pm$2.7 μm ($\pm$ SD).[68]

Usually the grafted serotonergic as well as noradrenergic cells are distributed in the host cord in two locations: a) within the solid tissue implants, b) as single cells within the host grey matter near the injection site (Fig. 2).[60,68] The question whether the transplantation of cell suspension rather than that of solid piece of tissue is more effective is still open. The former seems to hold greatest promise for extensive reinnervation of the larger areas of the cord due to a better chance for migration of the single cell from the injection site. However, it has been found recently that many serotonergic cells grafted in a solid piece of tissue migrated from the graft over a distance of up to 1.2 mm from the site of injection.[68] Moreover, it seems that there are certain advantages of a solid compared to a cell suspension grafts when inserted in a spinal cord cavity. First, the grafted solid tissue has an intact structure with the supporting glial and other cells intact and the cells suffer less damage from neuronal axotomy. Second, the solid graft is naturally less likely to be flushed away by cerebrospinal fluid at the time of surgery. Third, the solid transplant can be monitored following grafting by in vivo magnetic resonance imaging.[78]

It was also recently described that human solid grafts of embryonic spinal cord have a more restricted growth pattern compared to the grafted cell suspension.[78] Nevertheless, using either technique, the serotonergic or noradrenergic cells extended their axons into the distal part of the transected spinal cord, spreading rostrally up to the level of transection, and caudally for a considerable distance. In both cases the longest of the serotonergic axons reached the distance of 15-20 mm from the level at which serotonergic perikarya were still visible.[62,68,73] Thus, these results demonstrate that the effectiveness of reinnervation by serotonergic cells grafted into the spinal cord as suspension or as solid pieces of tissue was very similar.

The immunohistochemical observations were supported by using electron microscopy[72,79] where 5-HT-immunoreactive fibres from cells grafted as a suspension were selectively distributed in the ventral horn, interomediolateral cell column and in the dorsal horn. In all these

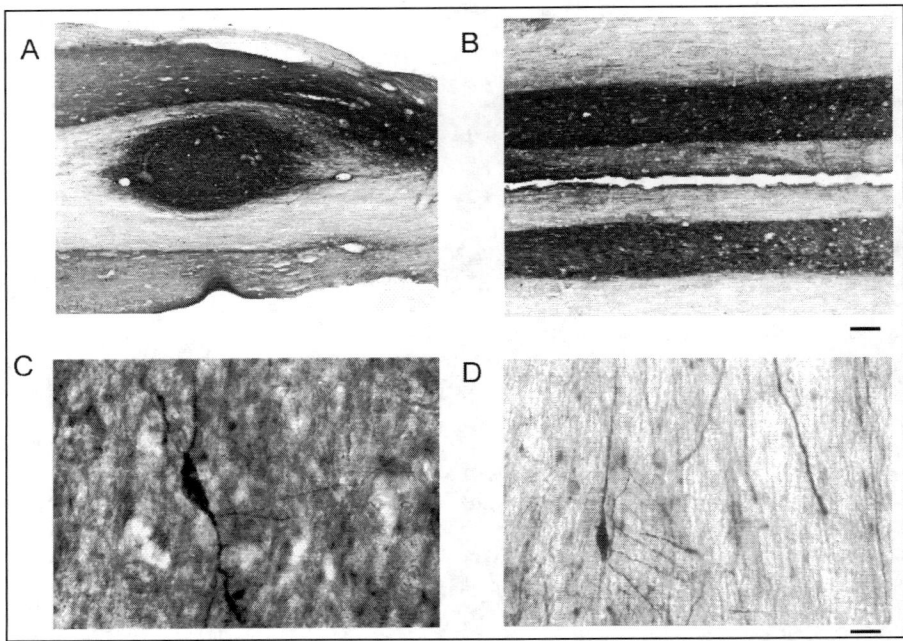

Figure 2. Photomicrographs of horizontal sections of the grafted rat spinal cord stained immunohistochemically for the presence of serotonin. A) Localisation of the 5-HT immunoreactive transplant in the dorsal part of the distal transected spinal cord. B) Distal part of the spinal cord at the level of central canal, few millimetres caudally from the transplant. Serotonergic neurons and their axons are visible in both the grey and white matter. C) High magnification photomicrograph of the grafted neurons and their axons. C) Stellate neurone in the grey matter of the distal part of the spinal cord in a grafted rat. D) Stellate neuron in the white matter of the spinal cord in the vicinity of the graft in a grafted rat. Calibration bar = 100 μm (A and B), and 20 μm (C and D) (modified from ref. 68).

regions they established conventional synaptic contacts, which were seen as numerous immunoreactive processes that contained microtubules, mitochondria and multivesicular bodies. These were postsynaptic to several terminal boutons. Thus, immunoreactive profiles of serotonergic fibres were found in the ventral horn that contains motoneurons, the superficial layers of the dorsal horn, and the intermedial column, which are the sites of 5-HT innervation in the intact animal.[73] Moreover, despite the use of rostral rhombencephalic raphe cells, which do not normally project to the spinal cord, a pattern of innervation similar to that seen in the intact spinal cord including specific innervation of motoneurons was seen.[72]

In some other experiments the distribution of spinal α_1-adrenoreceptors was investigated after grafting of the noradrenergic cell suspensions into the spinal cord below the total transection. It is well known that a complete transection of the spinal cord or chemical depletion of noradrenergic fibres in the spinal cord induces a significant increase of spinal α_1-adrenoreceptors.[80] The use of quantitative autoradiographic labelling of [3H]prazosin binding sites demonstrated that the transplantation of noradrenergic embryonic neurons from locus coeruleus reversed the lesion induced increase of spinal α_1-adrenoceptors in the grey matter of the damaged adult rat spinal cord.[74] Moreover, this reversal seems to be more effective after mechanical (spinal cord transection) than after chemical spinal cord denervation (5,7 DHT injection).

Whether the specific integration of transplanted monoaminergic cells with the neural tissue of the neural circuitry of the host spinal cord is related to the improvement of various spinal reflexes as well as to the level of recovery of the lost functions after spinal cord injury will be discussed in the next section.

Enhancement of Recovery of Function Following Intraspinal Grafting

After spinal cord injury, reflex and voluntary motor functions, as well as bladder and sexual reflexes, controlled by the central nervous system below the level of injury are initially lost. The recovery of functions mediated by supraspinally controlled reflexes is slow and varies with the severity of the injury and it's site. After incomplete spinal cord injury partial recovery of various functions occurs with time. As described earlier most of these functions are under monoaminergic control and their recovery depends on the level of monoamines present in the spinal cord below the lesion.

Since the properties of intraspinally grafted embryonic monoaminergic cells that survive develop in a donor-specific way, and establish synaptic connections with the host nervous system, they can replace lost nerve cells of the host. The first attempt to investigate the functional improvement in hindlimb movement recovery related to grafted noradrenergic cells was undertaken by Buchanan and Nornes in 1986.[60] In this study, cell suspension from embryonic brainstem, containing the locus coeruleus were injected into the lumbar spinal cord of adult rats. The hosts' spinal cord had been depleted of most descending catecholamine fibers by intracisternal injections of 6-OHDA. The grafted locus coeruleus neurons survived in the host cord, grew axonal processes of normal appearance that reinnervated the grey matter, and appeared to be functional since the hindlimb flexion reflexes were significantly enhanced. To assess the functional contribution of the transplanted noradrenergic cells, the force of the hindlimb flexion reflex was tested after acute spinal transection. It was shown previously that this reflex could be strongly enhanced by catecholamines,[65,81] and consistent with this in rats with grafted noradrenergic cells the flexion reflexes were significantly stronger than in the controls. Further evidence that the increase in the strength of the flexion reflex was mediated via noradrenergic inputs was provided by results where blocking α-adrenergic receptors abolished the enhancement of the flexion reflexes. These results demonstrated that the cells grafted from embryonic locus coeruleus survive, grow, extend catecholaminergic processes and affect the functional activity of the spinal cord.

The first investigation of functional improvement after grafting serotonergic cells into paraplegic rats (with the spinal cord totally transected at the thoracic level) was performed almost 15 years ago. Privat and coworkers[72] focussed their attention on the serotonergic control of sexual reflexes. In intact rats, following treatment with a specific inhibitor of 5-HT uptake (zimelidine), the incidence of ejaculation was increased, whereas erection was decreased.[39,82] In paraplegic rats treated with zimelidine, erection could be elicited but ejaculation did not occur. In contrast, in paraplegic rats that have received the graft of raphe cells stimulation of the penis was followed by ejaculation.[72] Thus, in addition to good survival of grafted cells, in paraplegic rats restoration of sexual reflexes also occurred.

Studies on the effects of raphe grafts on hindlimb locomotor abilities of paraplegic rats followed. Locomotor activity was tested in rats suspended over a moving treadmill or by examining fictive motor activities.[75,83] In these studies, eight days after injury in the region of thoracic cord, suspension of embryonic noradrenergic and/or serotonergic neurons was transplanted into the lumbar section of the spinal cord below the lesion. Results of the acute experiments performed 1 to 3 months later demonstrated that control paraplegic rats could exhibit fictive locomotion, but bilateral alternating activity recorded from motor nerveswas absent. In contrary, paraplegic rats that received grafts of embryonic locus coeruleus neurons bilateral, alternating, rhythmic locomotor-like activity could be recorded from motor nerves.

This effect was mediated by 5-HT containing neurons, since removal of noradrenergic cells by 6-hydroxydopamine had no effect on the restored locomotor-like activity. Furthermore, this locomotor like activity was facilitated when the reuptake of serotonin was blocked by zimelidine.[83] These results lead to the conclusions that reestablished serotonergic innervation was an important factor in the restoration of locomotor function of hindlimb muscles in adult paraplegic rats.

Surprisingly until recently there were few reports that evaluate thoroughly the degree and quality of functional motor recovery after total spinal cord transection with or without neural grafts in adult rats. A detailed study of the recovery of motor function in adult rats, with their spinal cord transected at the thoracic level, used several behavioral tests including kinematic and electromyographic analysis of hindlimb movement during bipedal locomotion[68,84] as well as some simple spinal reflexes.[68] The methods of grafting varied as well as the times after spinal cord transection were different. The main results of these studies confirm that grafted embryonic raphe nuclei that contain serotonergic cells are likely to encourage the recovery of hindlimb motor function in the adult paraplegic rats. The enhanced recovery of hindlimb motor functions is probably due to the increased excitability of the neuronal circuitry in the injured spinal cord. The greater excitability of the spinal neural circuitry in paraplegic rats that have received the graft of the embryonic brainstem containing raphe nuclei was indicated by the appearance of (a) the coordinated locomotor pattern, (b) the spontaneous air stepping, and c) the low threshold for evoking simple hindlimb reflexes.

The detailed study of locomotor like movements of hindlimbs, as well as several neurological tests revealed the main differences obtained from both experimental groups of paraplegic rats with or without a graft (Table 1). When the animals were observed during spontaneous behavior in their home cage without external stimulation it was noticed that in control paraplegic rats the hindlimbs were mainly passively extended with the ankle in plantar flexion without spontaneous movement - a typical picture of flaccid paralysis observed in paraplegic animals.[66,68] Only in very few animals without a graft were the hindlimbs kept in abduction with mild flexion of the ankle joint. In contrast, in most of the paraplegic rats that received a graft the hindlimbs were usually kept in flexion so that the plantar surface of the foot was touching the ground. The flexed position of hindlimbs in motionless sitting animals was obtained in about 65% of grafted rats. While in the paraplegic control rats the percentage of positive observation (flexed position of hindlimbs in animals sitting motionless) did not exceeded 10 % of the total number of observation. However, it is important to note here that in all paraplegic rats, grafted as well as control, the position of typical standing of intact rats with plantar surface of hindlimbs touching the ground (with clear weight body support) was never seen.

The tactile placing, i.e., the response to gentle tactile stimulation of the skin on the dorsal surface of the foot, was absent in all paraplegic rats, with or without the graft. In the intact animals the response usually consist of a lift (flexion) phase followed by an extension phase that provides support after contact with a surface. After spinal cord transection this response disappeared totally. However, in paraplegic rats that had a graft, the flexion phase (the dorsiflexion in the ankle joint) was present.

The next reflex activity tested was proprioceptive placing. The proprioceptive placing is an essential spinal response, which replaces the tactile placing in paraplegic animals. In paraplegic rats, which did not receive a graft passive displacement of the foot very rarely elicited a dorsiflexion of the ankle joint, while in rats which received a graft even very gentle touch of the dorsal part of the paw induced a slight change of the angle of the ankle joint and elicited the proprioceptive placing reaction which consisted of the dorsiflexion of the foot followed by relaxation or by withdrawal and extension towards the stimulus. Moreover, in some grafted rats an additional crossed reflex of the contralateral hindlimb was present.

Table 1. Comparing paraplegic rats with and without grafts

Neurological Test or Behavioral Observation	Paraplegic	Paraplegic with a Graft
Posture	No	No
Spontaneous hindlimb movement	No	No
Hindlimb flexion in sitting rats	No	Yes
Tactile placing	No	No
Proprioceptive tactile placing	Rarely	Yes
Spontaneous a r stepping	Rarely	Yes
Plantar walking with tail stimulation	No	Yes

The coordination of hindlimb during locomotor-like movements was also studied in paraplegic rats with their body supported on a small trolley which could be moved along a horizontal pathway[68] or when the animals were held over the moving treadmill.[84] In both experimental conditions, the paraplegic animals dragged theirs' hindlimbs behind the body over the moving ground or treadmill belt. Additional stimulation i.e., the continued pinching of the tail induced the hindlimbs of animals with a graft to carry out alternating movements with the pelvis lifted off the ground. None of the paraplegic control animals (without the graft) was able to do this. Figure 3 shows series of photographs taken during locomotion of (a) a control paraplegic rat without a graft, (b) paraplegic animal with a graft and (c) intact rat. The most important finding is that the grafted paraplegic rats could be induced to walk with regular alternating hindlimb movements and the plantar surface of their feet in contact with the ground during the stance phase, and ankle dorsiflexion during the swing phase of each step cycle. During such hindlimb locomotor-like movements of the paraplegic rats with a graft the stance and the swing phase alternated within each step cycle. In contrast, in the same experimental condition the control paraplegic animals (those with the spinal cord transected but without the graft) were not able to initiate the dorsiflexion of the ankle joint and dorsiflexion of the toes before the foot landed on the ground. Hindlimb movements in the paraplegic rats were produced mainly by their proximal joints (hip and knee) while the ankle joint movements were very limited, so that during any forward locomotion the dorsal surface of their feet dragged along the ground. With such limited movements of ankle joint it was impossible for the animals to support the body weight during the stance phase of the locomotor-like behavior. In contrast, in the grafted rats all three joints (ankle, knee and hip) of both hindlimbs were used during the locomotor-like movement and distinct plantar walking was obtained. However it is important to note that such hindlimb locomotor-like movement is no ideal and shows some signs of hyperextension when compared to normal hindlimb locomotion of intact rats (see Fig. 3B and C).

Air stepping was tested in the same animals.[68] Air stepping is a walking-like movement of hindlimbs commonly observed in animals after complete transection of the spinal cord when the rats are held off the ground in a vertical position. In most of our experiments, 3 months after spinal cord transection, the hindlimbs of the control paraplegic rats were flaccid when they were lifted off the ground to a vertical position. Tail pinching could elicit only brief episodes of small-amplitude movement of the hindlimbs, which were not alternating. The lifting off the ground of the grafted paraplegic rats to a vertical position very often elicited spontaneous episodes of air stepping lasting for 2-3 minutes. This 'walking-like' hindlimb movement without ground contact was more complex in the grafted rats and included vigorous movements of all the joints: hip, knee and ankle. While in the same situation in control rats only occasional uncoordinated movements of hindlimbs were obtained.

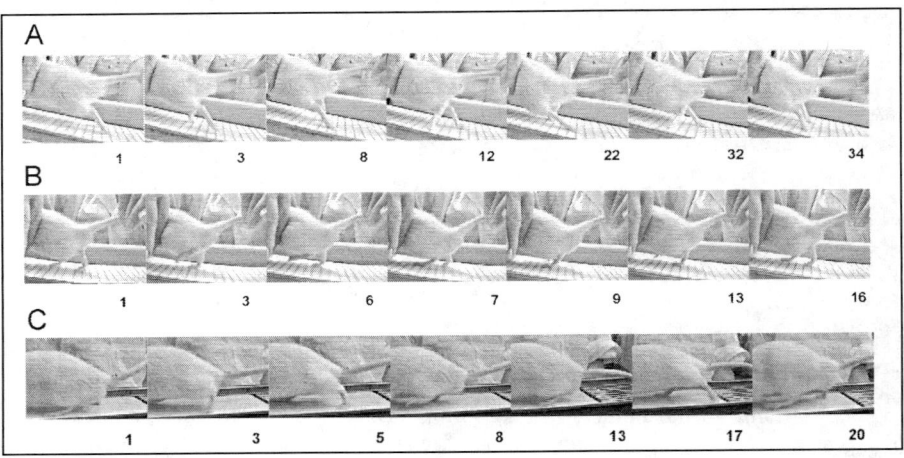

Figure 3. The series of photographs (the number of consecutive photo is displayed below each picture - 30 frames/s) taken during locomotor-like hindlimb movements 3 months after transection of the control rat with its spinal cord totally transected at the Th10 level (A); the rat with its spinal cord totally transected that has received the graft of embryonic raphe nuclei at the level of Th 12 (B) and regular locomotion on a treadmill of intact rat (C). Note the way of landing of the foot on the ground. (A) In the spinal control rats the locomotor like hindlimb movement is produced mainly by the proximal joints (hip and knee) while the dorsiflexion of the ankle joint is very limited. Note the proper plantar walking of the rat that had the spinal cord totally transected and received the graft of embryonic raphe region (B). (Modified from ref. 68).

Table 1 summarizes these results and provides evidence that the recovery of hindlimb locomotor functions after complete spinal cord transection is enhanced in rats that receive a solid graft of embryonic brainstem into the spinal cord below the lesion. In addition to improved restoration of hindlimb locomotor-like functions and restoration of several other reflexes (e.g., sexual reflexes, air stepping, tactile placing, proprioceptive reflexes) revealed that the spinal cord circuitry in paraplegic rats that received a graft of embryonic tissue was more responsive to afferent stimulation.

Summary

In the present chapter we described the use of neural tissue transplantation for enhancing the recovery of function that has been lost after injury of the spinal cord. Attempts to replace the descending monoaminergic inputs by embryonic grafts from the brain stem raphe nuclei to activate the remaining intact circuitry of the spinal cord below the lesion revealed that recovery of motor function of adult paraplegic rats was better when they received a graft containing serotonergic neurons. The behavioral experiments showed improvement in other spinal reflexes, too (sexual as well as tactile). Taken together these results confirm that the grafted cells of embryonic raphe nuclei, after integration with the host neural circuitry, are able to enhance recovery of hindlimb locomotor function. Thus the new monoaminergic connections between the grafted cells and host spinal cord provide a relatively precise targeting of serotonergic influences over the host spinal cord. The graft therefore acts not only as a source of a diffusely released neurotransmitter, but also can selectively activate or inhibit the appropriate cell populations of the host spinal cord.

References

1. Barbeau H, Rossignol S. Recovery of locomotion after chronic spinalization in the adult cat. Brain Res 1987; 412:84-95.
2. Barbeau H, Rossignol S. Initiation and modulation of the locomotor pattern in the adult chronic spinal cat by noradrenergic, serotonergic and dopaminergic drugs. Brain Res 1991; 546:250-260.
3. Barbeau H, Rossignol S. Enhancement of locomotor recovery following spinal cord injury. Curr Opin Neurol 1994; 7:517-524.
4. Hancock MB, Fougerousse CL. Spinal projections from nucleus locus coeruleus and nucleus subcoeruleus in the cat and monkey as demonstrated by the retrograde transport of horseradish peroxidase. Brain Res Bull 1976; 1:229-234.
5. Skagerberg G, Björklund A. Topographic principles in the spinal projections of serotonergic and nonserotonergic brainstem neurons in the rat. Neuroscience 1985; 15:445-480.
6. Fritschy JM, Grzanna R. Demonstration of 2 separate descending noradrenergic pathways to the rat spinal cord: Evidence for an intragriseal trajectory of locus coeruleus axons in the superficial layers of the dorsal horn. J Comp Neurol 1990; 291:553-582.
7. Bowker RM, Westlund KN, Sullivan MC et al. Organization of descending serotonergic projections to the spinal cord. Prog Brain Res 1982; 57:39-298.
8. Steinbusch HWM. Distribution of serotonin immunoreactivity in the central nervous system of the rat cell bodies and terminals. Neuroscience 1981; 6:557-618.
9. Mouchet P, Manier M, Feuerstein C. Immunohistochemical study of the catecholaminergic innervation of the spinal cord of the rat using specific antibodies against dopamine and noradrenaline. J Chem Neuroanatom 1992; 5:427-440.
10. Brodal A, Taber E, Walberg F. The raphe nuclei of the brain stem in the cat – II. Efferent connections. J Comp Neurol 1960; 114:239-259.
11. Dahlström A, Fuxe K. Evidence for the existence of monoamine-containing neurons in the central nervous system–I. Demonstration of monoamines in the cell bodies of brain stem neurons. Acta Physiol Scand Suppl 1964; 232:1-55.
12. Ahlman H, Grillner S, Udo M. The effect of 5-HTP on the static fusimotor activity and the tonic stretch reflex of an extensor muscle. Brain Res 1971; 27:393-396.
13. Ulfhake B, Arvidsson U, Cullheim S et al. An ultrastructural study of 5-hydroxytryptamine, thyrotropin-releasing hormone- and substance P-immunoreactive axon boutons in the motor nuclei of spinal cord segments L7-S1 in the adult cat. Neuroscience 1987; 23:917-929.
14. Holstage JC, Kuypers HGJM. Brainstem projections to spinal motoneurons: An update. Neuroscience 1987; 23:809-821.
15. Foote SL, Bloom FE, Aston-Jones G. Nucleus locus coeruleus: New evidence of anatomical and physiological specifity. Physiol Rev 1983; 63:844-914.
16. White SR, Neuman RS. Facilitation of spinal motoneurone excitability by 5-hydroxytryptamine and noradrenaline. Brain Res 1980; 188:119-127.
17. White SR, Neuman RS. Pharmacological antagonism of facilitatory but not inhibitory effects of serotonin and norepinephrine on excitabilty of spinal motoneurons. Neuropharmacology 1983; 22:489-494.
18. Jankowska E, Jukes MGM, Lund S et al. The effect of DOPA on the spinal cord. 5. Reciprocal organization of pathways transmitting excitatory actions to alpha motoneurons of flexors and extensors. Acta Physiol Scand 1967; 70:369-388.
19. Jankowska E, Jukes MGM, Lund S et al. The effect of DOPA on the spinal cord. 6. Half centre organization of interneurons transmitting effects from the flexor reflex afferents. Acta Physiol Scand 1967; 70:389-402.
20. Jankowska E, Hammar I, Djouhri L et al. Modulation of four types of feline ascending neurons by serotonin and noradrenaline. Eur J Neurosci 1997; 9:1375-1387.
21. Jankowska E, Gladden MH, Czarkowska-Bauch J. Modulation of responses of feline gama motoneurons by noradrenaline, tizanidine and clonidine. J Physiol (Lond) 1998; 512:521-531.
22. Jankowska E, Hammar I, Chojnicka B et al. Effects of monoamines on interneurons in four spinal reflex pathways from group I and/or group II muscle afferents. Eur J Neurosci 2000; 12:701-714.

23. Jankowska E, Riddel JS, Skoog B et al. Gating of transmission to motoneurons by stimuli applied in the locus coeruleus and raphenuclei of the cat. J Physiol (Lond) 1993; 461:705-722.
24. Grillner S. Locomotion in vertebrates—Central mechanisms and reflex interaction. Physiol Rev 1975; 55:247-304.
25. Grillner S, Zangger P. On the central generation of loccmotion in the low spinal cat. Exp Brain Res 1979; 34:241-261.
26. Shik ML, Orlovsky GN. Neurophysiology of locomotor automatism. Physiol Rev 1976; 56:465-500.
27. Wetzel MC, Stuart DG. Ensemble characteristics of cat locomotion and its neural control. Prog Neurobiol 1976; 7(1):1-98.
28. Cazalets JR, Borde M, Clarac F. Localization and organisation of the central pattern generator for hindlimb locomotion in newborn rat. J Neurosci 1995; 15:4943-4951.
29. Kjaerulff O, Kiehn O. Distribution of networks generating and coordinating locomotor activity in the neonatal rat spinal cord in vitro: A lesion study. J Neurosci 1996; 16:5777-5794.
30. Kremer E, Lev-Tov A. Localization of spinal network asspciated with generation of hindlimb loco-motion in the neonatal rat organization of its transverse coupling system. J Neurophysiol 1997; 77:1155-1170.
31. Fedirchuk B, Nielsen J, Petersen N et al. Pharmacologically evoked fictive motor patterns in the acutely spinalized marmoset monkey (Callitrhix jacuchus). Exp Brain Res 1998; 122:351-361.
32. Bussel B, Roby-Brami A, Yakovleff A et al. Late flexion reflex in paraplegic patients. Evidence for a spinal stepping generator Brain Res Bull 1989; 22:53-56.
33. Dimitrijevic MR, Gerasimenko Y, Pinter MM. Evidence for a spinal central pattern generator in humans. Ann N Y Acad Sci 1998; 860:360-376.
34. Grillner S. Control of locomotion in bipeds, tetrapods, and fish. In: Brookhart JM, Mountcastle VB, eds. Handbook of Physiology. The Nervous System, vol II, Motor Control, Am Physiol Soc, Bethesda, 1981:1179-1236.
35. Forssberg H, Grillner S. The locomotion of the acute cat injected with clonidine. i.v. Brain Res 1973; 50:184-186.
36. Björklund A, Skagerberg G. Descending projections to the spinal cord. In: Sjölund B, Björklund A, eds. Brainstem control of spinal mechanisms. Elsevier: Amsterdam, 1982:55-88.
37. Schroder HD. Organization of the motoneurons innervating the pelvic muscles of the male rat. J Comp Neurol 1980; 192:567-587.
38. Bancila M, Vergé D, Rampin O et al. 5-hydroxytryptamine2c receptors on spinal neurons control-ling penile erection in the rat. Neurosci 1999; 92:1523-1537.
39. Mas M, Zahradnik MA, Martino V et al. Stimulation of spinal serotonergic receptors facilitates seminal emission and suppresses penile erectile reflexes. Brain Res 1985; 342:128-134.
40. Marson L, List MS, McKenna KE. Lesions of the nucleus paragigantocelularis alter ex copula pe-nile reflexes. Brain Res 1992; 592:187-192.
41. Sachs BD, Garinello LD. Hypothetical spinal pacemaker regulating penile reflexes in rats: Evidence from transection of spinal cord and dorsal penile nerves. J Comp Physiol Psychol 1980; 94:530-535.
42. Tang Y, Rampin O, Calas A et al. Oxytocinergic and serotonergic innervation of identified lum-bosacral nuclei controlling penile erection in the male rat. Neuroscience 1998; 82:241-254.
43. Yells DP, Hendricks SE, Prendergast MA. Lesions of the nucleus paragigantocellularis: Effects on mating behavior in male rats. Brain Res 1992; 596:73-79.
44. Aloi JA, Insel TR, Mueller EA et al. Neuroendocrine and behavioral effects of m-chlorophenylpiperazine administration in rhesus monkeys. Life Sci 1984; 34:1325-1331.
45. Steers WD, DeGroat WC. Effects of m-chlorophenylpiperazine on penile and bladder function in rats. Am J Physiol 1989; 257:R1441-R1449.
46. Faden AI, Gannon A, Bausbaum AI. Use of serotonine immunocytochemistry as a marker of in-jury severity after spinal trauma in rats. Brain Res 1988; 450:94-100.
47. Shapiro S. Neurotransmission by neurons that use serotonine, noradrenaline, glutamate, glycine and γ-aminobutyric acid in the normal and injured spinal cord. Neurosurgery 1997; 40:168-176.
48. Commissiong JW. Fetal locus coeruleus transplanted into the transected spinal cord of the adult rat. Brain Res 1981; 271:174-179.
49. Hashimoto T, Fukuda N. Contribution of serotonin neurons to the functional recovery after spi-nal cord injury in rats. Brain Res 1991; 539(2):263-70.

50. Ellaway PH, Trott JR. The mode of action of 5-hydroxytryptophan in facilitating a stretch reflex in the spinal cat. Exp Brain Res 1975; 22:145-162.
51. Norman KE, Pépin A, Barbeau H. Effects of drugs on walking after spinal cord injury. Spinal Cord 1998; 36:699-715.
52. Barbeau H, Rossignol S. Enhancement of locomotor recovery following spinal cord injury. Curr opin Neurol 1994; 7:517-524.
53. Eriksson J, Olausson B, Jankowska E. Antispastic effects of L-dopa. Exp Brain Res 1996; 111:296-304.
54. Rémy-Néris O, Barbeau H, Daniel O et al. Effects of intratecal clonidine injection on spinal reflexes and human locomotion in incomplete paraplegic subjects. Exp Brain Res 1999; 129:433-440.
55. Yoshimura N, Sasa M, Ohno Y et al. Contraction of urinary bladder by central norepinephrine originating in the locus coeruleus. J Urol 1988; 139:423-427.
56. Morrison JFB, Spillane K. Neuropharmacological studies on descending inhibitory controls over micturition rflex. J Auton Nervous System Suppl 1986; 393-397.
57. Fung J, Stewart JE, Barbeau H. The combined effects of clonidine and cyproheptadine with interactive training on the modulation of locomotion in spinal cord injured subjects. J Neurologic Sciences 1990; 100:85-93.
58. Nygren LG, Olson L, Seiger A. Monoaminergic reinnervation of the transected spinal cord by homologous fetal grafts. Brain Res 1977; 129:225-235.
59. Björklund A, Nornes H, Gage FH et al. Reinnervation of the denervated spinal cord by grafted noradrenergic and serotonergic brain stem neurons. Development and Plasticity of the Mammalian Spinal Cord. In: Goldberger ME, Gono A, Murrray M, eds. Folia Research Series, vol. III, Section V. Regenerative Capacity of the Spinal Cord. Liviana Press, Padova, 1986:291-299.
60. Buchanan JT, Nornes HO. Transplants of embryonic brainstem containing the locus coeruleus into spinal cord enhance the hindlimb flexion reflex in adult rats. Brain Res 1986; 381:225-236.
61. Privat A, Mansour H, Pavy A et al. Transplantation of dissociated fetal serotonin neurons into the transected spinal cord of adult rats. Neurosci Lett 1986; 66:61-66.
62. Foster GA, Roberts MHT, Wilkinson LS et al. Structural and functional analysis of raphe neurone implants into denervated rat spinal cord. Brain Res Bull 1989; 22:131-137.
63. Foster GA, Roberts E, Gage FH et al. Restoration of function to the denervated spinal cord after implantation of embryonic 5HT- and substance 5 containing rephe neurons. Prog Brain Res 1990; 82:247-59.
64. Rajaofetra N, Köning N, Poulat P et al. Fate of B1-B2 and B3 rhombencephalic cells transplanted into the transected spinal cord of adults rats: Light and electron microscopic studies. Exp Neurol 1992; 117:59-70.
65. Nygren L-G, Olson L. On spinal noradrenaline receptor supersensitivity: Correlation between nerve terminal densities and flexor reflexes various times after intracisternal 6-hydroxydopamine. Brain Res 1976 116:455-470.
66. Feraboli-Lohnherr D, Orsal D, Yakovleff A et al. Recovery of locomotor activity in the adult chronic spinal rat after sublesional transplantation of embryonic nervous cells: Specific role of serotonergic neurons. Exp Brain Res 1997; 113:443-54.
67. Feraboli-Lohnherr D, Barthe J-Y, Orsal D. Serotonin-induced activation of the network for locomotion in adult spinal rats. J Neurosci Res 1999; 55:87-98.
68. Slawinska U, Majczynski H, Djavadian R. The recovery of hindlimb motor functions after spinal cord transection is enhanced by grafts of the embryonic raphe nuclei. Exp Brain Res 2000; 132:27-33.
69. Nornes H, Björklund A, Stenevi U. Reinnervation of the denervated adult spinal cord of rats by intraspinal transplants of embryonic brain stem neurons. Cell Tissue Res 1983; 230:15-35.
70. König N. Wilkie MB, Lauder JM. Tyrosin hydroxylase and serotonine containing cells in embryonic rat rhombencephalon: A whole-mount immunocytochemical study. J Neurosci Res 1988; 20:212-223.
71. Schmidt RH, Björklund A, Stenevi U. Intracerebral grafting of dissociated CNS tissue suspension: a new approach for neural transplatation to deep brain sites. Brain Res 1981; 218:347-356.

72. Privat A, Mansour H, Geffard M. Transplantation of fetal serotonin neurons into the transected spinal cord of adult rats: Morphological development and functional influence. Prog Brain Res 1988; 78:155-166.

73. Privat A, Mansour H, Rajaofetra N et al. Intraspinal transplants of serotonergic neurons in the adult rat. Brain Res Bull 1989; 22:123-129.

74. Roudet C, Gimenez y Ribotta M, Privat A et al. Intraspinal noradrenergic-rich implants reverse the increase of α_1 adreneceptors densities caused by complete spinal cord transection or selective chemical denervation: A quantitative autoradiographic study. Brain Res 1995; 677:1-12.

75. Yakovleff A, Roby-Brami A, Guezard B et al. Locomotion in rats transplanted with noradrenergic neurons. Brain Res Bull 1989; 22:115-121.

76. Björklund A, Segal M, Stenevi U. Functional reinnervation of rat hippocampus by locus coeruleus implants. Brain Res 1979; 170:409-426.

77. Björklund A, Stenevi U, Schmidt RH et al. Intracerebral grafting of neural cell suspension I. Introduction and general methods of preparation. Acta Physiol Scand Suppl 1983; 522:1-7.

78. Åkesson E, Holmberg L, Eriksdotter J et al. Solid human embryonic spinal cord xenografts in acute and chronic spinal cord cavities: A morphological and functional study. Exp Neurology 2001; 170:35-316.

79. Rajaofetra N, Ridet JL, Poulat P et al. Immunocytochemical mapping of noradrenergic projections to the spinal cord with an antiserum against noradrenaline. J Neurocytol 1992; 21:481-494.

80. Roudet C, Savasta M, Feuerstein C. Normal distribution of alpha-1 adrenoceptors in the rat spinal cord and its modification after noradrenegic denervation: A quantitative autoradiographic study. J Neurosci Res 1993; 34:44-53.

81. Grossman W, Jurna J, Nell T. The effect of reserpine and DOPA on reflex activity in the rat spinal cord. Exp Brain Res 1975; 22:351-361.

82. Blier P, De Montigny C. Electrophysiological investigation on the effect of repeated zimelidine administration on serotonergic neurotransmission in the rat. J Neurosci 1983; 3:1270-1278.

83. Yakovleff A, Cabelguen JM, Gimenez y Ribotta M et al. Fictive motor activities in adult chronic rats transplanted with embryonic brainstem neurons. Exp Brain Res 1995; 106:69-78.

84. Gimenez y Ribotta MG, Provencher J, Feraboli-Lohnherr D et al. Activation of locomotion in adult chronic spinal rats is achieved by transplantation of embryonic raphe cells reinnervating a precise lumbar level. J Neurosci 2000; 20:5144-5152.

CHAPTER 4

Encouraging Regeneration of Host Neurons:

The Use of Peripheral Nerve Bridges, Glial Cells or Biomaterials

Antal Nógrádi

Introduction

Recent results challenged the dogma that regeneration of CNS axons is impossible. These findings stimulated the interest of experimental neurobiologists and led to research that improved our understanding of the rules that control regeneration of structures in the mammalian spinal cord after injury or disease.

Therapeutic approaches can target either the acute management of the injured spinal cord to protect it from the secondary, autodestructive lesions, such as edema, inflammation, necrosis etc. and/or after the acute phase of the injury the functional restoration of the patients' lost functions.[1] This restoration can be at present attempted only by using functional methods or restorative neurology.

In experimental conditions of partial or complete spinal cord injury the restoration of lost spinal cord function was thought to be possible by using neuronal grafts to injured spinal cord to act as so-called "relay-bridges" or to release neurotransmitters (for details see Chapters 3 and 5). The evaluation of the effects of various cellular components taken from either the PNS or the CNS on regeneration are also discussed in this chapter.

Nonneuronal PNS elements and various materials have been used for decades to establish proper connections between lesioned and disconnected parts of the spinal cord and to restore the anatomical continuity and possibly the functional integrity between the separated elements of the cord. Most of these efforts simply wanted to build a "bridge" between the disconnected parts of the severed cord along which nerve fibres could pass and bridge the injured area. These experiments focused on encouraging the limited regenerative capacity of spinal cord neurons. Though initially there were some promising results obtained by several laboratories, the number of different approaches shows how desperate these attempts were to achieve any improvement in the sad outcome of a trauma to the spinal cord.

In this chapter we discuss the results of these attempts and their possible use for the management of patients with injured spinal cord.

Transplantation of Neural Tissue into the Spinal Cord, Second Edition,
edited by Antal Nógrádi. ©2006 Eurekah.com and Springer Science+Business Media.

Grafting Peripheral Nerves into the Spinal Cord

Intraspinal Nerve Bridges

The idea to use peripheral nerve segments as intraspinal grafts to improve regeneration in severed cords has its origin in the early decades of this century. Since Ramon y Cajal and Tello[2,3] proposed to use peripheral nerve bridges to promote axonal regeneration an increasing number of original articles have appeared often reporting fairly controversial effects of peripheral nerves implanted into transected spinal cords. The fact that this controversy could persist for several years was possibly due to inefficient functional analysis of spinal cord locomotion, possibly unsatisfactory surgical techniques, inadequate silver staining methods and rough electrical stimulation techniques in these studies.

The first pioneers of this promising grafting technique were Sugar and Gerard[4] whose classical experiments unfortunately could not be reproduced. They reported functional recovery after inserting degenerated sciatic nerve pieces in between the stumps of a transected cord. The recovered animals were reported to perform voluntary hindleg toe movements during walking and "peculiar hopping of both hindlegs alternated with stepping movements". Electrical stimulation of the brainstem elicited rhythmic stepping or extension of hindlimbs. Also, excellent correlation was claimed to exist between physiological recovery and anatomical restoration where newly formed axons connected the separated cord stumps via the grafted degenerated nerves. However, the source of these regenerating fibres was not established.

Later, other studies using the same grafting technique could neither reproduce nor confirm the above results. Barnard and Carpenter in 1950[5] used fresh or degenerated peripheral nerve auto- or homotransplants to reconnect the transected cord of rats. They observed no functional improvement in grafted animals though, morphologically ten days after grafting few axons penetrated the implanted nerve but none of these axons could be traced throughout the graft. Similarly negative results were reported by Brown and McCouch[6] as well as Feigin et al[7] who found no difference between grafted and control animals. Moreover, the anatomical analysis showed relatively poor fibre growth into the implant. Crude electrical stimulation of the brainstem and the cortex elicited either no or only a weak motor response in the hindlimbs in both control and grafted animals.[6,7] Accordingly, these authors claimed that no regeneration can occur through an intraspinal nerve bridge and concluded that Sugar and Gerard experiments were either based on incomplete spinal cord transection methods or misinterpreted (see also for earlier reviews: Windle 1956,[8] Nornes et al, 1984[9]).

These early experiments also suggested that the surgical technique used for either inflicting an injury or placing a graft into the spinal cord was very important. These studies all reported a connective tissue/glial scar formation at the nerve graft-spinal cord junction although its significance in the failure of regeneration was interpreted differently. Brown and McCouch[6] thought that this scar tissue was the impediment to axonal growth whilst Barnard and Carpenter[5] suggested on the basis of careful histological analysis that regeneration would not have been better in the absence of the scar. Later, electron microscopic and immunohistochemical investigations added significant amount of information towards elucidating this problem. It became evident that following transplantation of a peripheral nerve into the CNS the junction between the PNS and CNS tissue consisted of glial cells both of peripheral and central origin.[10] Astroglial cells separated the grafted nerve from the CNS environment but did not invade the graft significantly. On the contrary, Schwann cells from the nerve penetrated the host tissue for several millimeters and it was suggested that the axonal regeneration observed through the astroglial scar around the implanted nerve was probably due to the presence of Schwann cells.[10-12] Immunostaining to neurofilament proteins revealed that axonal regeneration was not confined to the implant but also occurred around the grafted tissue i.e., at sites where Schwann cells were present.[11] Thus it became clear that although glial and/or connective tissue scarring at the nerve-spinal cord junction is important but it is not solely responsible for the abortive regeneration and many other factors that may impede regeneration after such spinal cord injury have to be considered (for review see Guth et al, 1983[13]).

The histopathological reactions after spinal cord injury and following immediate grafting were analysed by Kao and his colleagues.[14-18] Their main observation was that transected spinal cords undergo a so-called "autotomy" which leads to the disruption of a segment of both the rostral and caudal stumps.[14] Transplantation of various nervous tissue such as nodose ganglion, sciatic nerve, cultured cerebral tissue[14] or spinal cord strips[19] has not prevented this necrotic process. Sometimes the graft itself disappeared several weeks after grafting or a cysts was formed at the graft-cord interface. Only peripheral nerve grafts were able to establish a structural continuity between the transected stumps but still necrotic changes developed in the transected stumps of the spinal cord rostral and caudal to the graft.[14] Further analysis of the pathological events after spinal cord transection, which was considered the least destructive method, revealed that the autolysis seen in the injured cord is chiefly due to lysosomal digestion of the separated cord stumps.[16-18] Kao et al concluded that spinal cord injury is followed by an imperfect wound healing resulting in initial microcyst formation, later cavitation and the necrotic tissue is replaced by a connective tissue scar. Electron microscopical investigations revealed that possibly regenerating axons form so-called terminal clubs as the result of abortive regeneration.[17] Accordingly, immediate grafting of a peripheral nerve segment between the transected stumps of a spinal cord which later undergoes partial necrosis will result in the degeneration of the graft and formation of dense collagenous scar between the implanted nerve and spinal cord.

Many of these problems were avoided by introducing more sophisticated surgical techniques and delayed grafting.[15,18,20] The introduction of a delicate transection method in dogs and rodents left the pia-arachnoid tubing intact and one week later, when the necrotic changes were complete, the necrotic tissue was carefully removed and grafting carried out.[18] The use of these procedures prevented to some extent the formation of connective tissue but not that of cavitation. Delayed grafting itself resulted in the formation of glial basement membrane at the stumps of the transected cord but this did not prevent some axonal regeneration across the graft-cord interface.[15] Although several months after grafting the grafted peripheral nerve contained both some thinly myelinated and unmyelinated regenerating axons the reinnervation of the graft still appeared poor. From these results it could be concluded that in the case of delayed grafting the main impediment to regeneration was the separation of the graft tissue from the host cord by the glial basement membrane. Later an important improvement has been achieved by Wrathall et al[21,22] who used cultured, nonneuronal peripheral cells (mainly Schwann cells) to fill the microscopic spaces between the contused host spinal cord and the grafted peripheral nerve. Thus the grafted tissue had several encouraging effects upon axonal growth into the graft. The cografted nonneuronal cells migrated both into the spinal cord and the peripheral nerve segment and induced a superior wound healing to that where only nerve segments were grafted.

Moreover, in cords where nonneuronal cells were cografted along with peripheral nerves the axonal growth was significantly promoted. Axons were observed in the grafted nerve as early as seven days after the delayed grafting and by this time these axons enwrapped by Schwann cells penetrated at least one mm into the nerve. However, the effect of these nonneuronal cells was temporary for one year after surgery no obvious difference could be observed between the reinnervation pattern of nerves grafted with or without nonneuronal peripheral cells.

Further factors that may be important for improvement of the axonal growth into peripheral nerve grafts have been explored recently. Senoo et al (1998) showed that grafts taken from the proximal stump of peripheral nerves ligated 7 days before grafting suppressed astrocytic scar formation at the host-graft interface and the number of regenerating axons was 10 time greater in the predegenerated grafts than in control animals with untreated grafts.[23] The glial environment of the host cord is also an important factor for regenerating axons. Sims et al (1999) found that host animals, whose spinal cord was x-ray irradiated at postnatal day 3, developed less complete astrocytic scar at the host-graft interface and Schwann cells intermingled with the host tissue. In nonirradiated animals many axon terminals, that traversed the peripheral nerve graft, ended blindly in the astrocytic scar.[24]

Notwithstanding the importance of the above attempts to restore the anatomical continuity of a transected or contused spinal cord these studies left several intriguing questions open. It was evident that some regenerating axons entered the intraspinally implanted nerve and possibly elongated till they reached the other end of the graft but the source of these fibres and their capacity to penetrate into the host spinal cord and establish synapses there remained unanswered. Most of the animals where evidence for morphologically successful regeneration into the graft was established were not analysed for recovery of function.

The quantitative analysis of reinnervation of intraspinal nerve grafts was performed by Richardson and his colleagues.[25-27] Injections of HRP into the transplanted spinal cord rostrally and caudally to the graft resulted in some retrograde labelling of spinal cord neurons in the vicinity of the other end of the graft.[25-27] However, the number of these neurons was relatively low suggesting that many fibres might have entered the graft but only few reentered the host spinal cord. According to combined autoradiographic and HRP retrograde labelling experiments the dorsal root ganglion cells that were close to the graft contributed significantly to the reinnervation of the transplanted nerve segment but not to that of the rostral spinal cord.[26] There was also a difference in the ability of axons of different origin to penetrate into the graft: axons of neurons near the implanted nerves entered the graft and some of these were able to reenter the spinal cord. The cell bodies were, however always a few mm away from the graft. Nerve segments bridging the brainstem and the cervical cord attracted regenerating neurites from more distant cell bodies: these grafts were reinnervated by neurons in the brainstem and thoracic cord as far as 2-3.5 cm rostral and 4-5 cm caudal from the nerve implant.[27] In contrast, neither bulbospinal nor corticospinal descending fibres invaded the graft suggesting that the long fibre tracts have a limited capacity to regenerate.[26]

Disappointingly, the grafts did not produce any obvious functional improvement or changes of primitive stepping movements. Considering the relatively low and varying number of neurons that were able to establish connections between the transected cord stumps via the nerve bridge this finding was not surprising particularly in view of the fact that none of the long descending pathways did enter the graft.[26]

Regeneration into grafts that replaced the dorsal columns of the cat spinal cord dorsal columns by a segment of the radial nerve was studied by Wilson.[28-30] It was suggested by the authors that partial injury of the spinal cord with preservation of the ventral half of the cord would not cause mechanical distraction and autolysis between the graft and the cord and better regeneration would occur into the graft as found in previous experiments. Although the presence of good nerve-graft junction and that of many axons traversing the nerve bridge was reported, HRP injections into the central portion of the graft resulted only in very low numbers of retrogradely labelled neurons in the spinal cord stumps and neighbouring dorsal root ganglia.[29,30] Evoked potentials could be recorded from the dorsal columns caudal and rostral to the graft in response to stimulation of the sciatic nerve demonstrating some degree of transmission through the graft. Along with these electrophysiological results good sensation, proprioception and muscular coordination were reported five months after grafting together with the regained ability to walk and use of the hind legs. Although this implantation technology tried to challenge the concept of "spinal cord autotomy" and improve regeneration into peripheral nerve grafts, unfortunately, in fact it did not achieve better results than reconnection of fully transected cords by others.

Genetic manipulation of the intraspinal implants has also been attempted to improve the axonal growth-promoting capacity of these grafts.[31,32] Intercostal nerve grafts, transfected with an adenovirus encoding neurotrophin-3, promoted the growth of more corticospinal axons through the nerves into the distal grey matter than in controls. The effect of NT-3 expression in Schwann cells on axonal growth also improved functional recovery in grafted animals.

One of the most interesting findings in this field was described by Cheng et al.[33] In their model the spinal cord of adult rats was completely transected and the proximal white matter bundles were connected by intercostal nerve grafts to the grey matter of the distal stump. The grafts were stabilized with fibrin glue containing acidic fibroblast growth factor. The neurotrophic and/or neuroprotective properties of this factor have not been fully revealed yet. Compared with untreated rats, animals with aFGF-peripheral nerve grafts showed regeneration of axons into the distal grey matter and better locomotor recovery of hindlimbs. Though this study reported functional connections between the transected spinal cord stumps, the rats showed limited locomotor recovery. Clinical and experimental studies revealed that remarkably few spinal axons are needed (cc 10%) for injured humans or animals to recover. This suggest that some minimal axonal connection between transected stumps may induce functional recovery.

Taken together, these experiments highlighted the fact that the concept of abortive regeneration can be challenged. Although it is evident that grafting a peripheral nerve into a severed spinal cord is neither morphologically nor functionally able to replace the damaged spinal cord segment but may induce significant regeneration of axotomized neurons of the host spinal cord and neighbouring PNS tissue (for more details see section: Extraspinal nerve bridges).

Extraspinal Nerve Bridges

The possibility of reconnecting separated parts of the central nervous system by using peripheral nerve grafts outside the CNS has been considered for a long time.[2,3,8] These ideas had the clear cut advantage that the grafted nerve was not affected by the degenerative events in the lesioned CNS and the ends of the graft could be placed into intact segments of the host spinal cord. Therefore their use offered an even more promising outcome and better regeneration than that of intraspinal grafts. In the early experiments the brain was the main target for such experiments and attempts were made to eliminate the glial barrier formed after a lesion and/or implantation of peripheral nerve grafts into the brain.[34] Initial attempts to bridge a complete lesion in the spinal cord were made few years later with surprisingly good outcome.[35,36] The transected spinal cord was bridged with a intercostal nerve one-two months after the first operation so that the ventral and dorsal roots of the nerve were left intact and its distal stump was implanted into the spinal cord beyond the lesion site. 25 dogs out of 30 were reported to have developed first signs of "reflex standing and walking" within 10-14 days after grafting. Moreover, resection of the bridge led to loss of the above functional improvements within two-three days. Although no morphological evidence of regenerating fibres traversing from the transected stumps via the scar tissue was found considerable number of regenerated fibres could be traced within the graft from the upper part of the spinal cord and the dorsal root ganglion.[36] In dogs which did not show any functional improvement very little axonal ingrowth was observed. It is difficult to evaluate these results, even with a significantly increased knowledge of 35 years after these pioneering experiments. The reconnection achieved by the ingrowing fibres was not quantified in these early experiments. Therefore it is difficult to accept in view of previously listed experiments with intraspinal nerve grafts[14-18,25-27] that such functional improvement was solely due to a relatively weak connection established by an intercostal nerve. Another possibility is that Schwann cells from the inserted nerve and from the lesion site invaded the injured spinal cord areas and promoted regeneration of intrinsic spinal cord axons and dorsal root afferents into the spinal cord. Ultrastructural studies by Lampert and Cressman[37] (1964) reported such ingrowth of axons along Schwann cells that invaded the lesioned cord and vessels. However, the growth of these regenerated fibres was not sustained for a sufficient length of time and they degenerated later. Unfortunately, this model of extraspinal nerve grafting was not used later so that the possible use this technique is unknown.

In another series of investigations initiated by Aguayo and his colleagues sciatic nerve segments were used to bridge segments of an intact spinal cord or connect the brainstem to certain parts of the cord.[38-40] This system had several advantages for the study of spinal cord regeneration.[38] The cord remained intact with minimal lesioning at the sites of grafting allowing long-term survival of the animals. Moreover, the origin and length of regenerated axons could also be documented using electrophysiological investigations. Nevertheless, no functional analysis could be performed as the grafting did not produce loss of neurological function. Nerve bridges connecting the medulla and the spinal cord contained regenerating fibres invading the nerves for several cm from both the cord and medulla as shown by retrograde labelling with HRP.[38,39] The distance of the cell bodies with regenerating axons from the graft appeared a very important factor. Only neurons in the vicinity of the graft and neighbouring dorsal root ganglia were able to grow their axons into the nerve implant.[40] Usually there were more retrogradely labelled cells caudally rather than rostrally to the graft and accordingly, the rostrocaudal extension of labelled neurons was greater caudally. In the cat, the rostral extension of regenerated axons was approximately 30 mm whilst caudally numerous labelled neurons were found when their axons were labelled, as far as 75 mm from the nerve implant.[41] On the other hand, despite this propensity of local spinal cord and dorsal root ganglion-derived neurons to grow into peripheral nerve grafts long tract fibres did not show such an ability to regenerate. Similarly to previous experiments on grafted Schwann cells (see section: Transplantation of Schwann cells) long descending axons rarely entered midthoracic or lumbar implants and ascending axons from the lumbar segments failed to regenerate into high cervical grafts.[40]

Electrophysiological recording from the peripheral nerves grafted into the medulla showed that the regenerated axons had the ability to propagate action potentials so that they originated from a functionally active CNS neurone.[42] Some of the axons had both spontaneous (synchronously active with the respiratory cycle) and induced (responding to sciatic nerve stimulation) activity which, at least in some cases resembled that of a normal cell present in the region of implantation. On the other hand, many axons in the graft remained silent. There was also evidence to suggest that some central neurons projecting into the graft responded to both excitatory and inhibitory transsynaptic influences. However, it was suggested that most neurons projecting into peripheral nerve grafts had reduced or altered synaptic input. Accordingly, it seemed conceivable that regenerated CNS neurons may be able to establish a simple neuronal circuitry in the lesioned CNS if a more complex experimental reconstruction is aimed.[42]

All these above studies raised a further question. What is the response of these regenerated central neurons to an injury of their axons in the grafted nerve? Normally only motoneurons or preganglionic sympathetic cells project their axons from the CNS into peripheral nerves and usually these axons have a long course in PNS environment. Lesioning a peripheral nerve results in rapid regeneration but is it true for any central neurone which just regenerated into a peripheral nerve?

In a series of experiments extraspinal nerve grafts bridging the medulla with the spinal cord were crushed close to the nerve-spinal cord junction 6-42 weeks after grafting, left to regenerate for further 4-11 weeks and labelled with HRP approximately 10 mm distally from the site of crush.[43] Retrograde labelling with HRP revealed a distribution and number of spinal cord cells similar to that seen in previous experiments using the same grafting and labelling procedure. Most of the labelled spinal cord neurons were intrinsic cells whose axons do not normally project into peripheral nerves. To prove the origin of regenerated fibres after crush injury, neurons injured by the crush were first labelled with Fast Blue and two weeks later the regenerated axons with Nuclear Yellow. The presence of double-labelled neurons (Fast Blue + Nuclear Yellow) suggested that neurons which regenerated into the peripheral nerve were able to regrow their injured processes and the regrowth was not due to collateral sprouting. This indicated that spinal cord neurons in general possess considerable regenerative capacity if a favourable axonal environment is provided. Nevertheless, in another series of experiments by Richardson and Verge[44] it became evident that regeneration and regrowth of central axons of otherwise peripheral dorsal root ganglion (DRG) neurons into a peripheral nerve graft is enhanced by a deleterious rather than a known encouraging event applied to the peripheral process. Thus, aggressive treatment with colchicine or section of the peripheral processes of DRG neurons enhanced regeneration of their central processes into the nerve graft while for example nerve growth factor had no such effect.[44]

Grafting peripheral nerves into a chronic spinal cord lesion was also able to induce recovery of certain populations of injured neurons.[45] Tibial nerves inserted into an aspiration cavity four weeks after the lesion induced many chronically injured neurons to extend their axons into the nerve graft. The distribution and size of cells was not uniform: from the neighbouring dorsal root ganglia mainly small cells regenerated whilst labelled spinal cord neurons originated from laminae III-VIII and X. This finding suggested that some chronically injured neurons may lose their potential to regenerate whilst others are able to retain it after a long delay and repeated injury. Recently it has been shown, that neurons of Clarke' nucleus showed increased survival when a peripheral nerve was grafted into the hemisected thoracic spinal cord.[46]

In view of the above results it can be concluded that spinal cord neurons have a tremendous potential to regenerate even after a repeated physical damage. These regenerating neurons are also functionally active, but have limited or altered synaptic contacts probably due to the absence of their target.[42]

Although this capacity of lesioned neurons to regenerate into a peripheral nerve is encouraging the functional reconstruction of severely lesioned spinal cords is probably not possible by reconnecting the lesioned segments with extraspinal peripheral nerve bridges. Instead, the regenerative capacity of the spinal cord neurons should probably be enhanced in other ways, using all the practical and theoretical experience accumulated in these studies.

Transplantation of Schwann Cells and Olfactory Ensheathing Cells into Lesioned Spinal Cord to Enhance Regeneration

Transplantation of Schwann Cells

Contrary to the rapid and effective regeneration in the PNS after a lesion to a peripheral nerve, injured axons in the CNS of adult mammals are not capable of considerable regeneration. Accordingly, the ability of axotomized spinal cord neurons to penetrate the lesion site in adult animals is limited but they are able to extend some neurites around the site of the lesion. This phenomenon is called "abortive regeneration". The causes for the abortive nature of the axonal growth are not understood. Nevertheless, there are few factors which have been considered to hinder axonal regeneration in the CNS. These include (a) the possible presence of inhibitory molecules on the surface of oligodendrocytes and myelin (b), glial scarring or (c) the lack of molecules stimulating axonal growth. Recent findings highlighted the fact that axotomized CNS neurons are able to grow processes, given the right external environment such as peripheral nerves implanted into the CNS (see Chapter 2). An increasing weight of evidence suggests that Schwann cells are the elements in the PNS which support axonal growth.[47] Accordingly, Kromer and Cornbrooks[48] reported considerable regeneration of the lesioned septohippocampal pathway after grafting cultured Schwann cells into the lesion (1985). Based on the success of these experiments Schwann cells were implanted into lesioned spinal cords with the possibility in mind that (a) Schwann cells may promote axonal regeneration of lesioned CNS nerve fibres by neutralizing the inhibitory effects present in the CNS and providing a favourable microenvironment for growing axons and (b) Schwann cell implants may cause a regression of the astrocytic gliosis at the site of the lesion.

In a number of studies Schwann cells were grafted into an injured cord where the injured portion of the spinal cord was free of axons and in the case of delayed transplantation astroglial-connective tissue mixed scar occupied the damaged part of the cord. The lesion was produced either photochemically,[49] by compressing the cord with an inflatable microballoon[50,51] or by creating a suction cavity in the corticospinal tract.[52] In spite of the different lesioning methods the outcome of Schwann cell grafting was very similar for each model. Moreover, the fact that the grafted Schwann cells were mixed in some studies with DRG cells[52,53] or collagen gel[49] did not seem to influence the effects of grafted Schwann cells on axonal regeneration. Syngeneic grafts of purified or mixed Schwann cells easily invaded the injured spinal cord and the area occupied by the grafted cells depended on the number of implanted cells. Cell-poor grafts produced only clusters of Schwann cell invasion intermingled with the remnants of scar tissue.[50] Larger number of grafted cells filled the lesion cavity completely and produced a very smooth interface between the host and grafted tissues.[50-52] The Schwann cell invasion was detected either by S-100 protein immunostaining or by prelabelling the grafted Schwann cells with an E.Coli galactosidase gene[51] so that the extent of repair by Schwann cells and the ingrowth of axons into the graft could be assessed.

The grafts had several effects, which could be taken to be "beneficial", on the lesioned cord. Immediate grafting after the injury resulted in more sprouting and ingrowth of host fibres (see later) but caused cyst formation in the grafted cord and the reduction of gliosis was only moderate.[52] Grafting performed between 2 and 4 days after injury led to poor survival of the grafted tissue probably because of release of cytotoxic factors in the lesioned cord.[51] Delayed grafting (five or more days after injury) improved the survival and the integration of the graft into the host spinal cord and often resulted in a very good fusion between implant and host cord with minimal glial scarring and no cyst formation. In order to achieve such results relatively large number of Schwann cells had to be implanted.[52] In these areas central glia, Schwann cells and myelin were intermingled.[52]

An important effect of grafted Schwann cells is, however, that they encourage the growth and regeneration of lesioned fibres of the host cord.[49-53] Martin et al[50,51] have reported ingrowth of peptidergic fibres into the Schwann cell clusters and their growth was probably strictly related to the presence of the implant because these axons were always accompanied by Schwann cells. Immunohistochemical stainings to neurotransmitters suggested that the majority of these fibres arose from dorsal root afferents and only a few of them were descending supraspinal afferents.[50,51] These latter rarely invaded the graft but ran along the margins of the implant. Paino and Bunge[49] using a silver stain detected ingrowth of myelinated and unmyelinated fibres into a Schwann cell-collagen implant as early as 14 days after grafting. The number of the invading axons increased with time and reached its maximum 28 days after implantation. These axons became myelinated by the grafted Schwann cells and often followed parallel paths within the graft. However, profuse axonal branching was characteristic only at the host-graft interface but not in the graft. Similar results were reported by Kuhlengel et al in 1990.[52,53] In their studies the corticospinal tract of neonatal rats was injured and the cavity filled with a mixed Schwann cell-neurone implant. Several months later the corticospinal tract was traced anterogradely with WGA-HRP injected into the motor cortex. Regenerated corticospinal fibres were found growing in fascicles along the border of the implant in the host grey matter but never in the graft. Immediate grafting after the injury improved the fasciculation and regeneration of the corticospinal tract whilst delayed grafting somehow resulted in lesser fasciculation. According to functional analysis implanted rats had no improved locomotion and functional recovery compared with lesioned control animals which also exhibited substantial recovery two weeks after injury. Human Schwann cells, as potential therapeutic cells in human CNS injuries, have also been used by Guest et al. Similarly to rat Schwann cell, human cells alone did not promote the regeneration of injured corticospinal axons,[54,55] while additional therapy with a monoclonal antibody (IN-1) raised against a myelin-associated, neurite growth inhibitory protein or with aFGF-fibrin glue supported regeneration of some fibres into the graft (Fig. 1).[55] These results indicated that Schwann cells alone are not able to alter the growth-inhibitory environment of the injured spinal cord.

Similar results were obtained with Schwann cells that were genetically modified to overexpress BDNF or NGF.[56,57] These Schwann cells showed similar phenotype as nontransfected cells and promoted the growth of more axons into the spinal stump distal to the trail of grafted cells than untreated Schwann cells.[57] Some axons from several supraspinal nuclei have also reached the distal portion of the cord. The presence of supraspinal axons in the grafts as well as in the distal stump of the spinal cord is encouraging for the establishment of locomotion.

Another factor influencing the effect of Schwann cells on regeneration was the presence of a demyelinating lesion. When both demyelination and axotomy were induced in the spinal cord, the Schwann cells spread 6-7 mm throughout the region of demyelisation where growth cones were also present.[58] Axotomy without demyelination did not induce such growth of severed axons. These findings suggest that Schwann cell transplantation combined with demyelinisation facilitates long distance axonal growth in the injured adult spinal cord.

Li and Raisman studied the effect of implanted Schwann cells in a minimal spinal cord lesion.[59] Small, circumscribed lesions in the corticospinal tract were filled with Schwann cells harvested from neonatal sciatic nerve so as to form a bolus. The Schwann cells induced sprouting of the lesioned fibres and some collateral fibres entered the superficial parts of the Schwann cell graft. Although this "minimal lesion" model is in no way comparable to human spinal cord injuries, it may suggest that Schwann cells may induce more regeneration where the secondary degenerative changes are also minimized.

Further attempts to establish a proper "bridge" between the stumps of the transected cord involved application of mini-guidance channels. Resorbable collagen rolls, poly-lactic acid or other polymers (e.g., PAN/PVC) seeded with purified Schwann cells were implanted into the lesioned cord.[60-64] Schwann cells survived in any of these channels and myelinated the ingrowing axons, but host astrocytes failed to enter the tubes. The ingrowing axons were mainly immunoreactive for CGRP, but not for monoaminergic transmitters.[61] Poly-lactic acid guidance channels induced the ingrowth of high number of axons from the host cord but after a peak at 2 month following grafting the fibre ingrowth and myelination decreased in these channels.[63] Bamber et al infused neurotrophins into the spinal cord caudal to the PAN/PVC channels seeded with Schwann cells.[64] They found that BDNF and NT-3 treatment did not increase the number of myelinated axons within the channel, but enhanced the ingrowth of these fibres into the host gray matter distal to the graft (Fig. 2). This effect of the above neurotrophins was not observed when the tube was filled with Matrigel, but had no Schwann cells inside.

On the basis of experiments with successful Schwann cell grafting it can be concluded that transplanted Schwann cells have several effects which do influence the restorative process of the injured host spinal cord. They reduce glial scarring and are able to enhance to some extent the otherwise abortive axonal growth and regeneration. Unfortunately, grafts of Schwann cells cannot form functioning bridges between the injured parts of the spinal cord since long descending fibres do not usually enter the graft. Nevertheless, they are able to render the

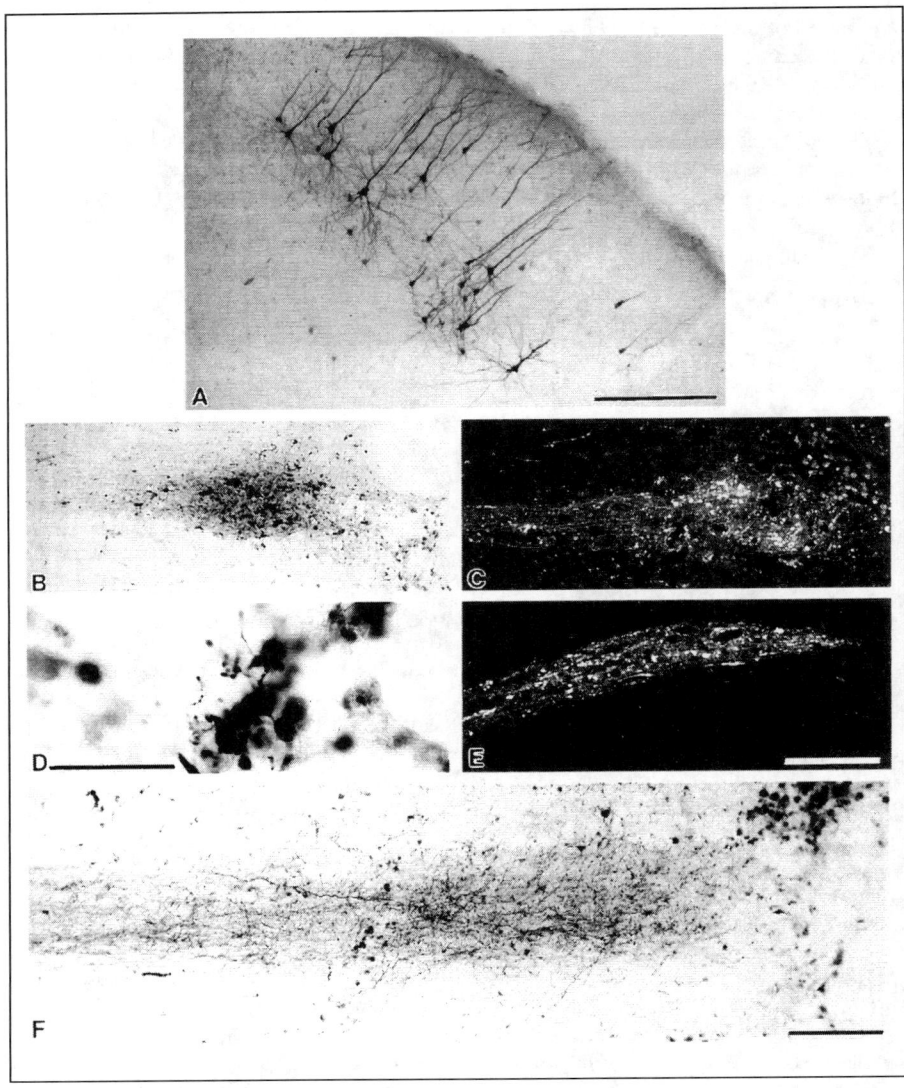

Figure 1. Corticospinal tract (CST) terminations in Schwann cell-grafted animals supplemented with aFGF-fibrin glue 35 days following graft implantation (rostral at right). A) A representative area of the cortex to demonstrate the strong labelling of neurons by biotinylated dextran amine (BDA). The terminations in B) (BDA-traced) and C) (FluoroRuby-traced) occured within 200 μm of the GFAP-defined interface. There are bulbous terminals but these are not found to be distributed rostrally over several millimeters as is observed with a Schwann cell graft only (E). In D) fine sprouts can be seen to arise from these BDA tracer-filled bulbous terminals in an animal treated with aFGF-fibrin glue. F) A marked degree of sprouting is seen over the distal 2 mm of this CST (aFGF-fibrin glue-treated). Scale bars = 0.5 mm in A; 0.5 mm in (B), (C), (E); 0.3 mm in (D); 0.2 mm in (F). Reproduced from Guest et al. J Neurosci Res 1997; 50:888-905, with kind permission from Pergamon Press Ltd, Oxford.

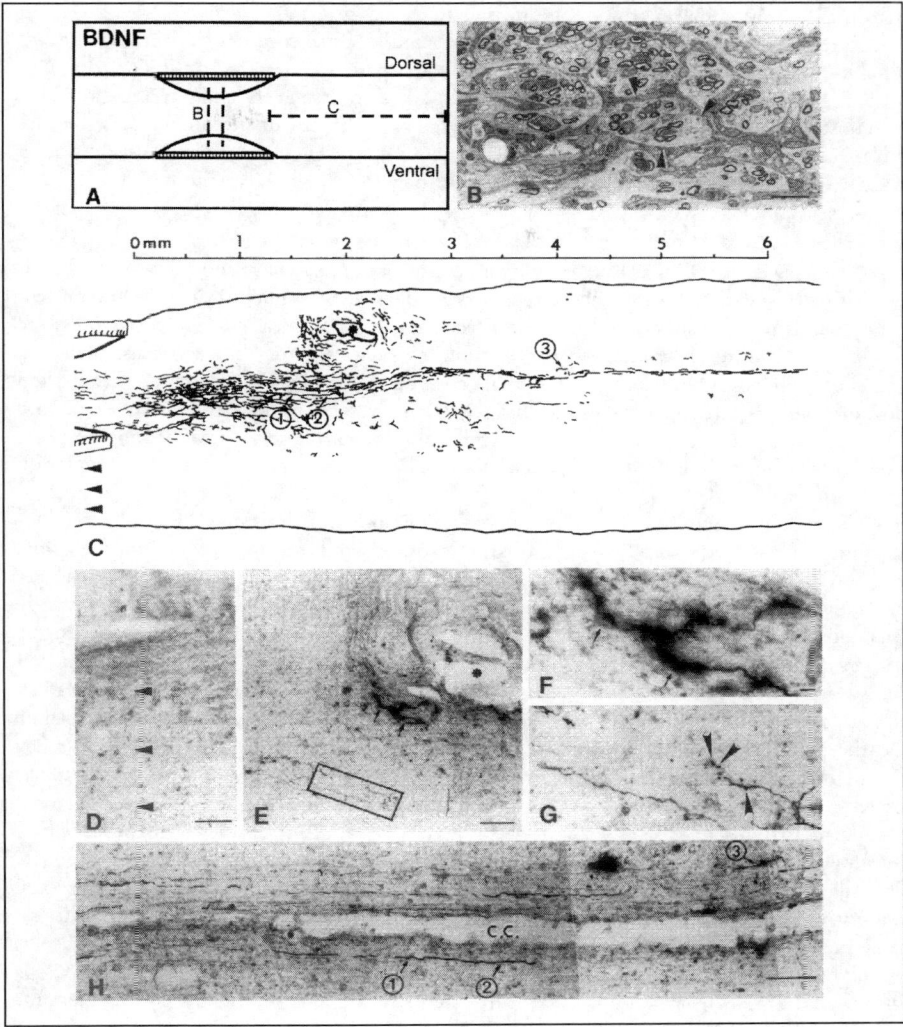

Figure 2. Axonal penetration into the distal host spinal cord after BDNF infusion. A) Schematic drawing showing the lateral view of the spinal cord and graft. Dashed lines indicate the positions of a cross-section of the graft (B) and a horizontal section of the distal graft and host cord (C). B) Toluidine blue-stained 1 μm plastic-embedded transverse section through the graft shows fascicles of myelinated axons (arrowheads). C) Camera lucida composite drawing, made from three horizontal sections, demonstrate PHA-L-labelled axons penetrating the distal graft-host interface and entering the spinal cord. Some labelled axons travelled for up to 6 mm. Some longitudinally running axons are indicated (1-3). Heavily labelled axonal profiles are found encircling the infusion site (asterisk). D) The complete transection of the intact left hemicord prevents the labelling of axons entering the distal cord via that route (arrowheads C and D). E) Many labelled axons (arrows) around the infusion site (asterisk). F-G) At higher magnification, labelled axons (F, arrows) and terminal bouton-like structures (G, arrowheads) are clearly seen. H) Labelled axons elongating in the vicinity of the central canal (c.c.) for a considerable distance. Bars, 10 μm (B,F and G); 100 μm (D,E, and H). Reproduced from Bamber et al. Eur J Neurosci 2001; 13:257-268, with kind permission from FENS/Blackwell Press.

surrounding host spinal cord more permissive to axonal growth. Grafting Schwann cells to stimulate regeneration as opposed to remyelination in the spinal cord has not resulted in functional improvement and their use for spinal cord injury patients is not yet applicable. Although purified and cultured Schwann cells can be maintained in tissue cell banks and are available from host peripheral nerves, they are still considered as alien elements in the spinal cord under normal circumstances and their long-term behaviour is not known. Their almost unlimited capacity to proliferate within a CNS environment may therefore be dangerous.

Transplantation of Olfactory Ensheathing Cells

Olfactory ensheathing cells (OECs) from the olfactory bulb have some common phenotypic properties with Schwann cells and astrocytes. They express glial fibrillary acidic protein (GFAP, an astrocyte marker) and form end-feet around blood vessels. They also express the low-affinity NGF receptor and produce laminin, characteristic for Schwann cells. However, they should be considered as a distinct glial cell type in the CNS. In the adult mammalian olfactory bulb, where olfactory ensheathing cells are present, normal and injured olfactory axons are able to elongate and establish synaptic contacts with other neurons throughout lifetime. These features indicated that grafted olfactory cells may be able to support regeneration in other parts of the CNS.

The pioneering studies of Ramon-Cueto and Nieto-Sampedro showed that olfactory ensheathing cells indeed promoted the growth of axotomised dorsal root axons into the spinal cord.[65] The dorsal roots were cut, attached to the dorsal surface of the cord and the gap between the root and the cord was bridged with olfactory glia cells. Ensheathing cells readily migrated into the spinal cord and were followed by regenerating dorsal root axons.[65] In the absence of OECs regenerating axons never reached the spinal cord suggesting that presence of these cells in the spinal cord was a prerequisite of successful regeneration. Spinal reflex restitution was studied in experiments where multiple lumbar (L3 to L6) rhizotomies were performed and the gap bridged with olfactory glial transplants.[66] The H reflex and withdrawal reflex returned by 60 days after grafting in most of the grafted animals while no recovery was observed in control animals.

Further studies followed to explore the growth-promoting effects of olfactory glia cells in the spinal cord. Li et al electrolytically injured the corticospinal tract and filled the lesion site with olfactory glia cells. Grafted animals showed extensive axonal growth and sprouting as well as improved forepaw reaching tasks compared with untreated animals. Further morphological studies revealed that the grafts contained at least two types of remyelinating olfactory glia cells, although their different role in the regenerative process is not clear.[67,68] The regenerating axons further penetrated the caudal host cord stump and later became myelinated by oligodendrocytes. This suggests that the environment around the lesion is permissive for axonal growth for a long time after injury. However, as in case of Schwann cell grafts, the "minimal lesioning" method introduced by Li et al did not allow to determine to what extent olfactory ensheathing cells enhance regeneration in a severely damaged spinal cord.

In another series of experiments Ramon-Cueto et al removed a 4 mm segment from the spinal cord, replaced it with a Schwann cell-seeded guidance channel and then injected olfactory glia suspensions into the proximal and distal spinal cord stumps.[69,70] OECs promoted the long-distance regeneration of both ascending and descending axons which readily entered the glial scars at the transected stumps and grew for several centimeters. Grafted olfactory glia cells migrated from the injection sites toward more rostral and caudal directions. When the transected cord stumps were injected with olfactory glia cells as above, but without Schwann cell-filled guidance channels, ensheathing cells induced significant functional recovery in paralysed animals. From 3 to 7 months after grafting, grafted animals supported their body weight, presented voluntary hindlimb movements and their hindlimbs responded to sensory stimuli. This functional improvement suggests that OECs may be the future choice of cells to be grafted into injured human spinal cords. However, the availability of purified olfactory glia cells raises more problems than that of Schwann cells. A potential source could be the nasal olfactory lamina propria of the individual which contains OECs.[71] This technology would make possible a kind of autologous transplantation provided that from the nasal mucosal surface, which is very small in humans, useful amount of tissue can be obtained. Alternatively, transgenic cells from large animals (such as pigs) can be used to obtain reasonable amount of tissue. Imaizumi et al[72] (2000) have shown that pig ensheathing cells genetically altered to reduce the hyperacute responses in humans are able to induce axonal elongation and restore impulse conduction in the transected spinal cord.

Implantation of Various Materials into the Spinal Cord

Implantation of Collagen

After early experiments had reported some success in reversing the abortive regeneration of lesioned spinal cord axons by grafting peripheral nerves, smooth muscle or fetal CNS cells some authors tried to use various biological materials, such as collagen gels, coated nitrocellulose or Millipore filters to form bridges between lesioned parts of the spinal cord and establish compatible environment for growing axons. Highly purified collagen matrices derived from animals appeared very promising for this purpose since recent results suggested that type I collagen inhibits the glial proliferation in vitro,[73] so that implanted collagen might decrease glial scar formation at the site of the lesion. Moreover, collagen matrices were reported to promote growth of severed central and peripheral axons and sustain the ingrowth of vascular elements.[74,75] On the basis of these results it seemed feasible that the use of a scaffolding structure placed into a damaged spinal cord could influence the cellular mechanisms of wound healing and facilitate axonal growth.

De la Torre[76,77] used first a cell-free bovine collagen matrix in a delayed grafting model. Collagen matrices implanted into the spinal cord 10 days after transection injury established a tight structural continuity between the transected ends and this natural protein integrated well with the host tissue. Newly formed blood vessels entered the implant and the new vessels were able to anastomose with the vascular supply of the spinal cord. Apart from new blood vessel formation within the implant, numerous and heterogenous cells invaded the biomatrix, including Schwann cells, macrophages, meningeal cells, fibroblasts and this latter actively produced new collagen.[76,78,79] However, in the first few days after implantation astrocytes rarely migrated into the collagen but later they penetrated into the biomatrix.[78,79] Microcyst formation (cavitation) was occasionally found in the bioimplant, mainly close to its proximal junction with the cord. De la Torre reported a limited axonal ingrowth of catecholaminergic fibres into the implant and some of these axons reentered the spinal cord.[76,77] Nevertheless, this limited regeneration did not cause any functional improvement, the treated animals did not regain sensory functions and showed no coordinated walking ability.

Other authors used slightly modified collagen gels but could not achieve better fibre ingrowth. Although there was observed only a moderate gliosis at the implant-host tissue interface[79] many axons did not enter the collagen but remained "dormant" at the interface. Gelderd[78] labelled the neurons projecting into the collagen matrix with HRP six weeks after implantation and found a higher number of labelled cells only rostral to the bioimplant than in control (transection only) animals.

Thus, several signs indicated that the structure of the collagen gel was of particular importance for wound healing and axonal regeneration. Gels treated with glyoxal upon implantation showed a good stability and dense texture formation whilst untreated gels became disorganized by massive infiltration of fibroblasts.[79] Paino and Bunge[49] used a different, three-dimensional collagen matrix which did not promote axonal growth at all and the authors suggested that this failure was possibly due to the unusual structure. More recently, Marchand et al[80] have reported an improved stability and durability of collagen implants by chemical cross-linking treatment. Cross-linked collagen gels have survived for at least six months after implantation and favoured regeneration because numerous axons extended into the implant and reentered the spinal cord.[80] Liu et al[81,82] used collagen guidance channels made of human placental collagen (type IV/IVoX) to bridge avulsed ventral roots and the spinal cord. In an elegantly designed experiment the collagen tube was implanted into the ventral horn of the spinal cord of marmosets and the avulsed C6 ventral root was inserted into the distal end of the tube. Collagen tubes alone or with an autologous peripheral nerve graft inside the tube induced the regeneration of numerous motoneurons into the denervated ventral root. Reinnervation by the growing axons of motoneurons induced considerable functional reinnervation in the biceps brachii muscle. It should be noted that the number of reinnervating motoneurons was nearly 7 times greater in the case of combined collagen tube-peripheral nerve implants compared with the use of collagen tubes alone.

However, the collagen gel implantation allowed further manipulations. Goldsmith and de la Torre[83] used in cats neurotrophic-like substances mixed in the gel before implantation into the transection gap and the implant was covered with an omental pedicle to improve blood supply of the biomatrix.[77] Two of the substances, 4-aminopyridine and laminin together with collagen matrix-omentum grafts induced tremendous axonal growth arising from descending monoaminergic tracts into the caudal stump as far as 90 mm below the lesion site. Retrograde labelling with Fluoro Gold showed labelled neurons in the brainstem nuclei which probably contributed to the reinnervation of the spinal cord. Consistent with these morphological findings coordinated forehindlimb locomotion was reported in cats treated with 4-aminopyridine or laminin.

Although, these findings are encouraging and have considerable clinical implications, they should be treated with caution until further studies will be conducted to determine the usefulness of these biomatrices.[83]

Implantation of Other Biosynthetic Materials

Apart from the studies on implantation of collagen biomatrix into injured spinal cord several other attempts have been made to bridge lesion cavities or rather just promote axonal regeneration by using bioimplants. Nitrocellulose treated with biological substances is known to support neurite growth in vitro.[84] Similarly, when nitrocellulose implants were placed into the spinal cord of newborn rats[84] to obstruct the growth of the corticospinal tract, the growing fibres which reached the implant several days (six-eight days) after implantation were able to grow and penetrate the implant. Successful regeneration was reported only in cases where the nitrocellulose filter was coated with laminin or had been kept previously in vitro with cultured spinal cord tissue for three-four days. Implantation of untreated nitrocellulose or exposure of the filter to cultures of cerebral cortex or spinal cord tissues longer than eight days before grafting did not support axonal growth of the corticospinal tract. The preliminary success with the coated nitrocellulose was considered to be due to the presence of certain adhesion molecules which are not present in cerebral or mature spinal cord cultured cells. Houlé et al[85-87] implanted a strip of nitrocellulose paper treated with nerve growth factor (NGF) in conjunction with a fetal spinal cord tissue transplant in order to enhance ingrowth of injured dorsal root fibres into the cord. They reported nearly three times enhanced axonal outgrowth compared to controls with untreated nitrocellulose implants. Interestingly, it was observed that the regenerating fibres were separated from the NGF-treated implant by a continuous layer of macrophages and astrocytes whilst untreated nitrocellulose was surrounded only with scattered macrophages and some astrocytes.[86] Thus the effect of NGF on axonal growth must have been indirect, probably mediated by the nonneuronal cell lining of the implant. Regenerating axons readily entered the rostral host spinal cord indicating that ascending axons were primarily promoted to grow across the injury site.[87] However, it could be that a combination of structural and biochemical effects mediated by astrocytes derived from the fetal spinal cord and other nonneuronal cells associated with the NGF-treated implant was responsible for the enhanced regeneration of dorsal root axons.

A similar strategy using a Millipore filter implant coated with embryonic spinal cord astrocytes was used[88-90] to determine whether the regrowth of sensory axons from injured dorsal roots into the implanted spinal cord can be improved. Such implants promoted the growth of crushed dorsal root fibres into the adult cord and inhibited glial scar formation, but axons leaving the nearby regions of the implant failed to enter the surrounding white matter. Retrograde labelling of regenerating fibres with HRP revealed that many of them formed axonal terminals with boutons suggesting some degree of synaptic plasticity. Although all the above studies confirmed the active function of immature astrocytes coating nitrocellulose or Millipore implants, it remained to be determined how they interact with other cell types, such as macrophages and Schwann cells in order to promote regeneration and what the molecular mechanism of their action is.[86,89]

Recent studies involved the use of a number of new materials, such as PHPMA (NeuroGel™) or polyHEMA hydrogels,[91,92] poly(D,L-lactide) foams modified by PELA copolymer, fibronectin mats and carbon filaments.[93-95] The hydrogels and poly(D,L-lactide) foam, when implanted into injured spinal cords, promoted axonal growth into the implant, angiogenesis and cell migration. Carbon filaments reportedly directed the growth of axons and migration of astrocytes into the bridge formed by the filaments.

In another series of studies the axonal growth-promoting effect of freeze-dried alginate has been studied.[96-98] Alginate is a bioabsorbable long chain polysaccharide, isolated from brown seeweed. Suzuki and colleagues developed a novel freeze-dried alginate gel to for spinal cord implantation. Alginate gel was implanted in the gap between the stumps of the transected cords of infant and adult rats. Axons penetrated the gel implants which became infiltrated with vessels, glial cells and some macrophages, although no signs of inflammation were observed. Some axons have been remyelinated. The regenerating axons left the alginate gels and entered the host cord: axonal elongation into the host cord could be followed for 1-1.5 cm rostrally and 200-300 μm caudally (Fig. 3). The morphological reinnervation of the host cord stumps was accompanied by electric activity resulting from the axons regenerating through the alginate implant and moderate functional recovery. These results suggest that alginate is a promising biomaterial in spinal cord regeneration studies.

Catecholaminergic neurons in rat mesencephalon were shown by using histofluorescence techniques to exhibit a considerable capacity to regenerate by growing and sprouting[99] following electrolytic lesions. Based on these results, Björklund et al used iris and mitral valve grafts for studying the regeneration of descending 5-HT and catecholaminergic fibres in the spinal cord.[100] The grafts placed into a compressed spinal cord became invaded by catecholaminergic fibres and formed loose, irregular plexuses within the grafted tissue. In contrast to this finding, 5-HT fibres showed abundant sprouting around the graft but were not seen to grow into it. The pattern of reinnervating fibres, in particular in the mitral valve, often resembled that of the original innervation of these tissues. The regenerated fibres probably followed the pathways presented by peripheral glial cells in the denervated iris or mitral valve.[101] However, the regeneration of these monoaminergic fibres did not occur only in the presence of the graft. Newly formed axons were able to penetrate the necrotic tissue that was formed in the absence of the graft after the compression injury.

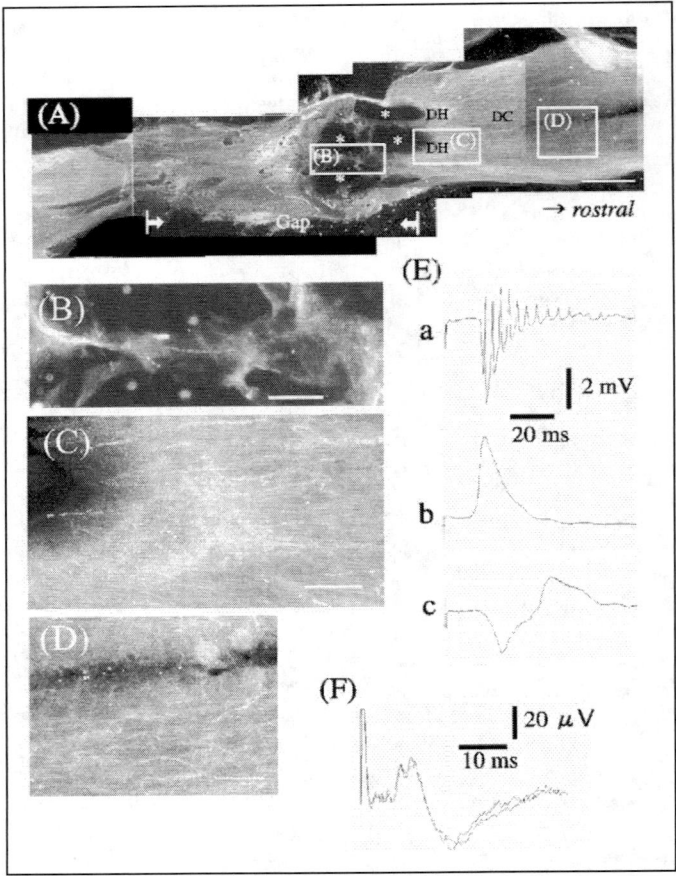

Figure 3. A-D) Darkfield photomicrographs of horizontally sectioned spinal cord. These micrographs show that HRP-labelled regenerating fibres from the dorsal funiculus grew massively through the alginate-implanted gap, reentered the rostral side, and extended randomly through the spinal cord tissue. A) Overall view of alginate-implanted gap and adjacent rostral as well as caudal areas. Numerous HRP-labelled regenerating axons are found within the gap. There are some cystic cavities (*) in the rostral part of the gap. The alginate sponge has completely disappeared. The rectangular areas are enlarged in (B-D). Scale bar, 1 mm. DC, dorsal column; DH, dorsal horn. B) Higher magnification of boxed area in (A). Regenerating axons are seen growing through the cystic wall rostral in the gap. Scale bar, 200 μm. C) Higher magnification of boxed area in (A). Labelled regenerating axons extend through host white and grey matter. The regenerating axons are not confined to specific spinal nerve tracts. Scale bar, 200 μm. D) Higher magnification of boxed area in (A). Many HRP-labelled regenerating axons are seen in the dorsal column rostral to the gap. Scale bar, 200 μm. E,F) Electrical potentials recorded in GN with glass microelectrodes. AC recording. E) Three responses (a, b, c) to stimulation of the sciatic nerve in three GN neurons at a depth of 0.5-1.0 mm from the dorsal surface of GN. a) The stimulus evoked repetitive spike responses recorded extracellularly from a single GN neuron. b) EPSP recorded intracellularly from another neuron. The falling phase of EPSP is curtailed by IPSP. c) IPSP followed by a longer latency EPSP. F) Field potentials recorded extracellularly at the dorsal surface of the medullary apex evoked by stimulation of the sciatic nerve. Time constant of recording system is 0.5 s in all these records. Upward deflection shows positivity in all records. Reproduced from Y. Suzuki et al, Neuroscience Letters 2002; 318:121-124, with kind permission from Elsevier Science Ireland Ltd.

Other tissues such as nodose ganglion and smooth muscle have also been grafted into the transected spinal cord without any particular success.[14]

These studies have shown that the biological materials or various nonneural tissues themselves did not induce significant regeneration when transplanted into the lesioned spinal cord. The most important finding is, however, that the cotransplanted nonneuronal cells, mainly immature astrocytes made the environment around the implant more permissive for axonal regeneration. Such role for immature glial cells is proposed elsewhere in this book (see Chapter 2) and discussed in details.

Transplantation of Genetically Modified Fibroblasts

Apart from exogenous Schwann cells or biomaterials carrying factors that may encourage the growth of host axons, genetically modified fibroblast have also been used for this purpose. It is relatively easy to culture fibroblasts and efficiently transduce them with adeno- or retroviral vectors and they have several advantageous features, such as low immunogenecity as autografts and minimal risk of tumor formation that make them appropriate for ex vivo gene transfer.[102] Grafting of genetically modified fibroblasts expressing various molecules (NGF, BDNF, neural cell adhesion molecule L1, etc) into an injured spinal cord[103-105] resulted in reliable synthesis of the molecules which were thought to enhance regeneration. Behavioral and morphological analysis showed that the use of genetically modified fibroblasts accelerated recovery from spinal cord injury[104] and/or promoted regeneration of injured axons.[103,105] However, these experiments also suggested that although fibroblasts possess features that make them candidates of cells to be grafted into an injured spinal cord, immunosuppressive therapy is needed to enhance the survival of grafts of genetically modified fibroblasts.

Application of Omental Tissue to Injured Spinal Cord

Goldsmith et al[106] suggested first in 1975 that the use of transposed omental grafts onto the surface of injured spinal cord would improve the autodestructive processes in the cord. It was known well before these studies that omental grafts applied to heart or brain can increase the blood flow in the recipient tissues. Therefore, application of omental tissue to lesioned cord was expected to influence the progressive fall in spinal cord blood flow at the site of injury by adding a new source of blood supply within a reasonable time. Indeed, omental tissue placed onto the dorsal surface of intact or lesioned spinal cords initiated revascularization of the cord within three days, i.e., omental vessels anastomosed with those of the spinal cord.[106,107] Moreover, the application of omentum reportedly diminished the edema at the site of injury within 24 hours and in some cats functional recovery was reported one month after surgery. However, these changes could be observed only when the omentum was applied to the cord immediately following the injury. Obviously, such rapid surgical treatment could not occur in a clinical situation. Recently, there are reports of neurological improvements in human patients following omental transposition in chronic spinal cord injuries.[108] Although the efficacy of this procedure was doubtful, great number of successful operations were claimed. In fact, two prospective studies of the efficacy of omental transposition in humans have shown that this procedure either failed to improve the morphological outcome in spinal cord injury or the neurological scores became even slightly worse after omental transposition.[109,110] In light of these thorough studies it can be argued that this procedure cannot be a suggested protocol in patients suffering from chronic spinal cord injury.

References

1. Lindsay KW, Bone I, Callander R. Neurology and Neurosurgery illustrated. Edinburgh: Churchill Livingstone, 1991.
2. Cajal SLY. Degeneration and Regeneration of the Nervous System. May RM, ed. London: Oxford University Press, 1928.
3. Tello F. La influencia del neurotropismo en la regeneración de los centros nerviosos. Trab Lab Invest Univ Madrid 1911; 9:123-159.
4. Sugar O, Gerard RW. Spinal cord regeneration in the rat. J Neurophysiol 1940; 3:1-19.
5. Barnard JW, Carpenter W. Lack of regeneration in spinal cord of rat. J Neurophysiol 1950; 13:223-228.
6. Brown JO, McCouch GP. Abortive regeneration of the transected spinal cord. J Comp Neurol 1947; 87:131-137.
7. Feigin I, Geller EH, Wolf A. Absence of regeneration in the spinal cord of the young rat. J Neuropath Exp Neurol 1951; 10:420-425.
8. Windle WF. Regeneration of axons in the vertebrate central nervous system. Physiol Rev 1956; 36:427-440.
9. Nornes H, Björklund A, Stenevi U. Transplantation strategies in spinal cord regeneration. In: Sladek JR, Gash DM, eds. Neural Transplants Development and Function. New York, London: Plenum Press, 1984:407-421.
10. Weinberg EL, Raine CS. Reinnervation of peripheral nerve segments implanted into the rat central nervous system. Brain Research 1980; 198:1-11.
11. Chi NH, Bignami A, Bich NT et al. Autologous sciatic nerve grafts to the rat spinal cord: Immunofluorescence studies with neurofilament and gliofilament (GFA) antisera. Exp Neurol 1980; 68:568-580.
12. Matsuyama Y, Mimatsu K, Sugimuru T et al. Reinnervation of peripheral nerve segments implanted into hemisected spinal cord estimated by transgenic mice. Paraplegia 1995; 33:381-386.
13. Guth L, Reier PJ, Barrett CP et al. Repair of the mammalian spinal cord. TINS 1983; 20-24.
14. Kao CC. Comparison of healing process in transected spinal cords grafted with autogenous brain tissue, sciatic nerve, and nodose ganglion. Exp Neurol 1974; 44:424-439.
15. Kao CC, Chang LW, Bloodworth Jr JMB. Axonal regeneration across transected mammalian spinal cords: An electron microscopic study of delayed microsurgical nerve grafting. Exp Neurol 1977; 54:591-615.
16. Kao CC, Chang LW. The mechanism of spinal cord cavitation following spinal cord transection. Part 1: A correlated histochemical study. J Neurosurg 1977; 46:197-209.
17. Kao CC, Chang LW, Bloodworth Jr JMB. The mechanism of spinal cord cavitation following spinal cord transection. Part 2: Electron microscopic observations. J Neurosurg 1977; 46:745-756.
18. Kao CC, Chang LW, Bloodworth Jr JMB. The mechanism of spinal cord cavitation following spinal cord transection. Part 3: Delayed grafting with and without spinal cord retransection. J Neurosurg 1977; 46:757-766.
19. Bunge RP, Johnson MI, Thuline D. Spinal cord reconstruction using cultured embryonic spinal cord strips. In: Kao CC, Bunge RP, Reier PJ, eds. Spinal Cord Reconstruction. New York: Raven Press, 1983:341-358.
20. Derlon JM, Roy-Camille RR, Lechevalier B et al. Delayed spinal cord anastomosis. In: Kao CC, Bunge RP, Reier PJ, eds. Spinal Cord Reconstruction. New York: Raven Press, 1983:223-234.
21. Wrathall JR, Rigamonti DD, Braford MR et al. Reconstruction of the contused cat spinal cord by the delayed nerve graft technique and cultured peripheral nonneuronal cells. Acta Neuropathol 1982; 57:59-69.
22. Wrathall JR, Kao CC, Rigamonti DD et al. Preparation of large quantities of nonneuronal cells from peripheral nervous tissue for spinal cord reconstruction. In: Kao CC, Bunge RP, Reier PJ, eds. Spinal Cord Reconstruction. New York: Raven Press, 1983:317-325.
23. Senoo E, Tamaki N, Fujimoto et al. Effects of peripheral nerve grafts on nerve regeneration in the rat spinal cord. Neurosurgery 1998; 42:1347-1356.
24. Sims TJ, Durgun MB, Gilmore SA. Transplantation of sciatic nerve segments into normal and glia-depleted spinal cords. Exp Brain Res 1999; 125:495-501.

25. Richardson PM, McGuinness UM, Aguayo AJ. Axons from CNS neurones regenerate into PNS grafts. Nature 1980; 284:264-265.
26. Richardson PM, McGuinness UM, Aguayo AJ. Peripheral nerve autografts to the rat spinal cord: Studies with axonal tracing methods. Brain Research 1982; 237:147-162.
27. Richardson PM, Aguayo AJ, McGuinness UM. Role of sheath cells in axonal regeneration. In: Kao CC, Bunge RP, Reier PJ, eds. Spinal Cord Reconstruction. New York: Raven Press, 1983:293-304.
28. Wilson DH. Peripheral nerve implants in the spinal cord in experimental animals. Paraplegia 1984; 22:230-237.
29. Wilson DH. Anatomical and physiological assessments of peripheral nerve grafts in the dorsal columns of the spinal cord. Rest Neurol and Neurosci 1991; 2:251-254.
30. Wardrope J, Wilson DH. Peripheral nerve grafting in the spinal cord: A histological and electrophysiological study. Paraplegia 1986; 24:370-378.
31. Blits B, Dijkhuizen PA, Carlstedt TP et al. Adenoviral vector-mediated expression of a foreign gene in peripheral nerve tissue bridges implanted in the injured peripheral and central nervous system. Exp Neurol 1999; 160:256-267.
32. Blits B, Dijkhuizen PA, Boer GJ et al. Intercostal nerve implants transduced with an adenoviral vector encoding neurotrophin 3 promote regrowth of injured rat corticospinal tract fibres and improve hindlimb function. Exp Neurol 2000; 164:25-37.
33. Cheng H, Cao Y, Olson L. Spinal cord repair in adult paraplegic rats: Partial restoration of hindlimb function. Science 1996; 273:510-513.
34. Windle WF, Clemente CD, Chambers WW. Inhibition of formation of a glial barrier as a means of permitting a peripheral nerve to grow into the brain. J Comp Neurol 1952; 96:359-369.
35. Perkins L, Babbini A, Freeman LW. Distal-proximal nerve implants in spinal cord transection. Neurology 1964; 14:949-954.
36. Turbes CC, Freeman LW. Peripheral nerve-spinal cord anastomosis for experimental cord transection. Neurology 1958; 8:857-861.
37. Lampert P, Cressman M. Axonal regeneration in the dorsal columns of the spinal cord of adult rats: An electron microscopic study. Lab Invest 1964; 13:825-839.
38. David S, Aguayo AJ. Axonal elongation into peripheral nervous system "bridges" after central nervous system injury in adult rats. Science 1981; 214:931-933.
39. Aguayo AJ, David S, Bray GM. Influences of the glial environment on the elongation of axons after injury: Transplantation studies in adult rodents. J Exp Biol 1981; 95:231-240.
40. Richardson PM, Issa VMK, Aguayo AJ. Regeneration of long spinal axons in the rat. J Neurocytol 1984; 13:165-182.
41. Sceats DJ Jr, Friedman WA, Sypert GW et al. Regeneration in peripheral nerve grafts to the cat spinal cord. Brain Research 1986; 362:149-156.
42. Munz M, Rasminsky M, Aguayo AJ et al. Functional activity of rat brainstem neurons regenerating axons along peripheral nerve grafts. Brain Research 1985; 340:115-125.
43. David S, Aguayo AJ. Axonal regeneration after crush injury of rat central nervous system fibres innervating peripheral nerve grafts. J Neurocytol 1985; 14:1-12.
44. Richardson PM, Verge VMK. The induction of a regenerative propensity in sensory neurons following peripheral axonal injury. J Neurocytol 1986; 15:585-594.
45. Houle JD. Demonstration of the potential for chronically injured neurons to regenerate axons into intraspinal peripheral nerve grafts. Exp Neurol 1991; 113:1-9.
46. Yick LW, Wu W, So KF et al. Peripheral nerve grafts and neurotrophic factors enhance neuronal survival and expression of nitric oxide synthase in Clarke's nucleus after hemisection of the spinal cord in adult rat. Exp Neurol 1999; 159:131-138.
47. Kleitman N, Wood P, Johnson MI et al. Schwann cell surfaces but not extracellular matrix organized by Schwann cells support neurite outgrowth from embryonic rat retina. J Neurosci 1988; 8:653-663.
48. Kromer LF, Cornbrooks CJ. Transplants of Schwann cell cultures promote axonal regeneration in the adult mammalian brain. Proc Natl Acad Sci USA 1985; 82:6330-6334.
49. Paino CL, Bunge MB. Induction of axon growth into Schwann cell implants grafted into lesioned adult spinal cord. Exp Neurol 1991; 114:254-257.

50. Martin D, Schoenen J, Delree P. et al. Grafts of syngeneic, adult dorsal root ganglion-derived Schwann cells to the injured spinal cord of adult rats: Preliminary morphological studies. Neurosci Lett 1991; 124:44-48.

51. Martin D, Schoenen J, Delree P. et al. Syngeneic grafting of adult rat DRG-derived Schwann cells to the injured spinal cord. Brain Res Bull 1993; 30:507-514.

52. Kuhlengel KR, Bunge MB, Bunge RP et al. Implantation of cultured sensory neurons and Schwann cells into lesioned neonatal rat spinal cord. IInd ed. Implant characteristics and examination of corticospinal tract growth. J Comp Neurol 1990b; 293:74-91.

53. Kuhlengel KR, Bunge MB, Bunge RP. Implantation of cultured sensory neurons and Schwann cells into lesioned neonatal rat spinal cord. Ist ed. Methods for preparing implants from dissociated cells. J Comp Neurol 1990a; 293:63-73.

54. Guest JD, Bunge RP. Functional studies of human Schwann cells transplanted to the nude rat spinal cord. J Neurotrauma 1995; 12:427.

55. Guest JD, Hesse D, Schnell L et al. Influence of IN-1 antibody and acidic FGF-fibrin glue on the response of injured corticospinal tract axons to human Schwann cell grafts. J Neurosci Res 1997; 50:888-905.

56. Tuszynski MH, Weidner N, McCormack M et al. Grafts of genetically modified Schwann cells to the spinal cord: Survival, axon growth, and myelination. Cell transplant 1998; 7:187-196.

57. Menei P, Montero-Menei C, Whittemore SR et al. Schwann cell genetically modified to secrete human BDNF promote enhanced axonal growth across transected adult rat spinal cord. Eur J Neurosci 1998; 10:607-621.

58. Keirstead HS, Morgan SV, Wilby MJ et al. Enhanced axonal regeneration following combined demyelisation plus Schwann cell transplantation therapy in the injured adult spinal cord. Exp Neurol 1999; 159:225-236.

59. Li Y, Raisman G. Schwann cells induce sprouting in motor and sensory axons in the adult spinal cord. J Neurosci 1994; 14:4050-4053.

60. Paino CL, Fernandez-Valle C, Bates ML et al. Regrowth of axons in lesioned adult spinal cord: Promotion by implants of cultured Schwann cells. J Neurocytol 1994; 23:433-452.

61. Xu XM, Guénard V, Kleitman N et al. Axonal regeneration into Schwann cell-seeded guidance channels grafted into transected adult rat spinal cord. J Comp Neurol 1995; 351:145-160.

62. Xu XM, Zhang S-X, Li H et al. Regrowth of axons into the distal spinal cord through a Schwann cell-seeded mini-channels implanted into hemisected adult rat spinal cord. Eur J Neurosci 1999; 11:1723-1740.

63. Oudega M, Gautiér SE, Chapon P et al. Axonal regeneration into Schwann cell grafts within resorbable poly(α-hydroxyacid) guidance channels in the adult rat spinal cord. Biomaterials 2001; 22:1125-1136.

64. Bamber NI, Li H, Lu X et al. Neurotrophins BDNF and NT-3 promote axonal reentry into the distal host spinal cord htrough Schwann cell-seeded mini-channels. Eur J Neurosci 2001; 13:257-268.

65. Ramón-Cueto A, Nieto-Sampedro M. Regeneration into the spinal cord of transected dorsal root axons is promoted by ensheathing glia transplants. Exp Neurol 1994; 127:232-244.

66. Navarro X, Valero A, Gudiòo G et al. Ensheathing glia transplants promote dorsal root regeneration and spinal reflex restitution after multiple lumbar rhizotomy. Ann Neurol 1999; 45:207-215.

67. Li Y, Field PM, Raisman G. Repair of adult rat corticospinal tract by transplants of olfactory ensheathing cells. Science 1997; 277:2000-2002.

68. Li Y, Field PM, Raisman G. Regeneration of adult rat corticospinal axons induced by transplanted olfactory ensheathing cells. J Neurosci 1998; 18:10514-10524.

69. Ramón-Cueto A, Plant GW, Avila J et al. Long-distance axonal regeneration in the transected adult rat spinal cord is promoted by olfactory ensheathing glia transplants. J Neurosci 1998; 18:3803-3815.

70. Ramón-Cueto A, Cordero MI, Santos-Benito FF et al. Functional recovery of paraplegic rats and motor axon regeneration in their spinal cords by olfactory ensheathing glia. Neuron 2000; 25:425-235.

71. Lu J, Féron F, Ho SM et al. Transplantation of nasal olfactory tissue promotes partial recovery in paraplegic adult rats. Brain Res 2001; 889:344-357.

72. Imaizumi T, Lankford KL, Burton WV et al. Xenotransplantation of transgenic pig olfactory ensheathing cells promotes axonal regeneration in rat spinal cord. Nat Biotechnol 2000; 18:949-953.

73. Eccleston PA, Mirsky R, Jessen KR. Type I collagen preparations inhibit DNA synthesis in glial cells of the peripheral nervous system. Exp Cell Res 1989; 182:173-185.

74. Madison R. Sidman RL, Nyilas E. et al. Nontoxic nerve guides support neovascular growth in transected rat optic nerve. Exp Neurol 1984; 86:448-461.

75. Madison R, Da Silva CF, Dikkes P. et al. Increased rate of peripheral nerve regeneration using bioresorbable nerve guides and a laminin-containing gel. Exp Neurol 1985; 88:767-772.

76. de la Torre JC. Catecholamine fiber regeneration across a collagen bioimplant after spinal cord transection. Brain Research Bulletin 1982; 9:545-552.

77. de la Torre JC, Goldsmith HS. Increased blood flow enhances axonal regeneration after spinal cord transection. Neurosci Lett 1988; 269-273.

78. Gelderd JB. Evaluation of blood vessel and neurite growth into a collagen matrix placed within a surgically created gap in rat spinal cord. Brain Research 1990; 511:80-92.

79. Marchand R, Woerly S. Transected spinal cords grafted with in situ self-assembled collagen matrices. Neurosci 1990; 36:45-60.

80. Marchand R, Woerly S, Bertrand L et al. Evaluation of two cross-linked collagen gels implanted in the transected spinal cord. Brain Res Bullet in 1993; 30:415-422.

81. Liu S, Bodjarian N, Langlois O et al. Axonal regrowth through a collagen guidance channel bridging spinal cord to the avulsed C6 roots: Functional recovery in primates with brachial plexus injury. J Neurosci Res 1998; 51:723-734.

82. Liu S, Said G, Tadie M. Regrowth of the rostral spinal axons into the caudal ventral roots through a collagen tube implanted into hemisected adult rat spinal cord. Neurosurgery 2001; 49:143-151.

83. Goldsmith HS, de la Torre JC. Axonal regeneration after spinal cord transection and reconstruction. Brain Research 1992; 589:217-224.

84. Schreyer DJ, Jones EG. Growth of corticospinal axons on prosthetic substrates introduced into the spinal cord of neonatal rats. Dev Brain Res 1987; 35:291-299.

85. Houlé JD. Johnson JE. Nerve growth factor (NGF)-treated nitrocellulose enhances and directs the regeneration of adult rat dorsal root axons through intraspinal neural tissue transplants. Neurosci Lett 1989; 103:17-23.

86. Houlé JD. Regeneration of dorsal root axons is related to specific nonneuronal cells lining NGF-treated intraspinal nitrocellulose implants. Exp Neurol 1992; 118:133-142.

87. Houlé JD, Ziegler MK. Bridging a complete transection lesion of adult rat spinal cord with growth factor-treated nitrocellulose implant. J Neurol Transpl Plast 1994; 5:115-124.

88. Kliot M, Smith GM, Siegal J et al. Induced regeneration of dorsal root fibres into the adult mammalian spinal cord. In: Reier PJ, Bunge RP, Seil FJ, eds. Current Issues in Neural Regeneration Research. New York: Alan R Liss Inc, 1988:311-328.

89. Kliot M, Smith GM, Siegal JD et al. Astrocyte-polymer implants promote regeneration of dorsal root fibres into the adult mammalian spinal cord. Exp Neurol 1990; 109:57-69.

90. Inoue HK, Kobayashi S, Ohbayashi K. Regeneration of hemisectioned spinal cord with and without supporting materials. Neurol Med Chir (Tokyo) 1997; 37:600-605.

91. Woerly S, Pinet E, de Robertis L et al. Spinal cord repair with PHPMA hydrogel containing RGD peptides (NeuroGel™). Biomaterials 2001; 22:1095-1111.

92. Giannetti S, Lauretti L, Fernandez E, et al. Acrilic hydrogel implants after spinal cord lesion in the adult rat. Neurol Res 2001; 23:405-409.

93. Priestley JV, Ramer MS, King VR et al. Stimulating regeneration in the damaged spinal cord. J Physiol (Paris) 2002; 96:123-133.

94. Maquet V, Martin D, Scholtes F et al. Poly(D,L-lactide) foams modified by poly(ethylene-oxide)-block-poly(D,L-lactide) copolymers and aFGF: In vitro and in vivo evaluation for spinal cord regeneration. Biomaterials 2001; 22:1137-1146.

95. Neelima CB, Figlewicz HM, Khan T. Carbon filaments direct the growth of postlesional plastic axons after spinal cord injury. Int J Dev Neurosci 1999, 17:255-264.

96. Suzuki K, Suzuki Y, Ohnishi K et al. Regeneration of transected spinal cord in young adult rats using freeze-dried alginate gel. Neuroreport 1999; 10:2891-2894.

97. Kataoka K, Suzuki Y, Kitada M et al. Alginate, a bioresorbable materila derived from brown seaweed, enhances elongation of amputated axons of spinal cord in infant rats. J Biomed Mater Res 2001; 54:373-384.

98. Suzuki Y, Kitaura M, Wu S et al. Electrophysiological and horseradish peroxidase-tracing studies of nerve regeneration through alginate-filled gap in adult rat spinal cord. Neurosci Lett 2002; 318:121-124.

99. Katzman R, Björklund A, Owman CH, et al. Evidence for regenerative axon sprouting of central catecholamine neurons in the rat mesencephalon following electrolytic lesions. Brain Res 1971; 25:579-596.

100. Björklund H, Dahl D, Olson L et al. Glial fibrillary acidic protein-like immunoreactivity in the iris: Development, distribution, and reactive changes following transplantation. J Neurosci 1984; 4:978-988.

101. Björklund A, Katzman R, Stenevi U et al. Development and growth of axonal sprouts from noradrenaline and 5-hydroxytryptamine neurones in the rat spinal cord. Brain Research 1971; 31:21-33.

102. Liu Y, Himes BT, Tyron B et al. Intraspinal grafting of fibroblasts genetically modified by recombinant adenoviruses. Neuroreport 1998; 9:1075-1079.

103. Kobayashi S, Miura M, Asou H et al. Grafts of genetically modified fibroblasts expressing neural cell adhesion molecule L1 into transected spinal cord of adult rats. Neurosci Lett 1995; 188:191-194.

104. Kim DH, Gutin PH, Noble LJ et al. Treatmentwith genetically engineered fibroblasts producing NGF or BDNF can accelerate recovery from traumatic injury in the adult rat. Neuroreport 1996; 7:2221-2225.

105. Liu Y, Kim D, Himes BT et al. Transplants of fibroblasts genetically modified to express BDNF promote regeneration of adult rat rubrospinal axons and recovery of forelimb function. J Neurosci 1999; 19:4370-4387.

106. Goldsmith HS, Duckett S, Chen WF. Spinal cord vascularization by intact omentum. Am J Surg 1975; 129:262-265.

107. Goldsmith HS, Steward E, Chen WF, et al. Application of intact omentum to the normal and traumatized spinal cord. In: Kao CC, Bunge RP, Reier PJ, eds. Spinal Cord Reconstruction. New York: Raven Press, 1983:235-243.

108. Rafael H. Omental transposition and spinal cord injury. J Neurosurg 1997; 87:800.

109. Clifton GL, Donovan WH, Dimitrijevic MM et al. Omental transposition in chronic spinal cord injury 1996; 34:193-203.

110. Duffill J, Buckley J, Lang D et al. Prospective study of omental transposition in patients with chronic spinal cord injury. J Neurol Neurosurg Psychiatry 2001; 71:73-80.

CHAPTER 5

Encouraging Regeneration of Host Neurons:

Transplantation of Neural Tissues into the Injured Spinal Cord Grafts of Embryonic Neural Tissue*

Gerta Vrbová

Introduction

Traumatic injury to the spinal cord causes disruption of the long descending and ascending pathways, degeneration of neurons in the lesioned area and destruction of the intrinsic spinal connections. The natural history of spinal cord injury involves formation of a dense gliotic scar surrounding the lesion site and diffuse astrocytic gliosis. The degree of neurological impairment depends on the extent of the lesion and, as discussed in Chapter 2, the species and age. A variety of neurological syndromes may evolve. In the mammals after total spinal cord transection para- or tetraplegia develops, with loss of sphincter control and sensation below the lesioned level. This sad outcome illustrates how limited, and for practical purposes nonexistent, the regenerative capacity of the mammalian spinal cord is. For decades many experimental efforts have focused on the problem of regeneration and functional recovery in the spinal cord. As discussed in the previous chapter, the initial experimental approach was to "bridge" the lesion site with different nonneuronal tissues, or various materials, in order to create a conducive environment for axonal regeneration. The outcome of these studies was encouraging; a limited axonal growth into, and sometimes even through, the lesion site did occur, showing that appropriate experimental manipulations could overcome the apparent lack of regenerative capacity of the CNS axons. However, this limited axonal growth had no impact on the impaired function of the animal. The neurobiologists in the field are still searching for the Holy Grail of a method of spinal cord repair.

In the late 1970s a new technique of repair of damaged neuronal circuitries by transplantation of embryonic neural tissues was introduced, and soon became explored in studies on regeneration, plasticity and functional recovery in the CNS. Studies in different experimental models of brain injury demonstrated that the grafts could indeed innervate and restore neurotransmitter release in the denervated areas of the host brain, encourage axonal growth and reverse some functional deficits (reviewed by Björklund and Stenevi, 1984;[1] Dunnett and Björklund, 1987;[2] Lindvall and Björklund, 1990;[3] Anderson et al, 1995).[4] With the advent of neurotransplantation the research into spinal cord repair took a new direction.

---

*Based on chapter in the previous edition written by Katarzyna Sieradzan.

Transplantation of Neural Tissue into the Spinal Cord, Second Edition,
edited by Antal Nógrádi. ©2006 Eurekah.com and Springer Science+Business Media.

Several lines of interest soon emerged. The first one naturally followed the previous work on the use of implants of nonneuronal tissues to encourage axonal growth. The use of embryonic neural tissue transplanted into the lesion site in the host spinal cord introduced a new element to these experiments. It was conceivable that, in contrast to other tissues or materials, the neurons contained in such grafts could engage in the process of reciprocal synapse formation with the neurons of the host, both by receiving innervation from the ingrowing local axons and by sending off their own axons into the rostral and caudal stumps of the severed spinal cord (for review see Bregman et al, 2002).[5]

According to this hypothesis, the graft could provide a "relay bridge" for the host axons, and not just a passive substrate for their elongation (Fig. 1). Both embryonic homotopic i.e., spinal cord, and heterotopic, for example cortical, cerebellar, dorsal root ganglia and hippocampal grafts, have been shown to survive in the spinal cord. However, in the majority of studies homotopic grafts have been used.

An approach based on the restoration of neurotransmitter release in the spinal cord caudal to the lesion has been exclusively studied. In the spinal cord the monoaminergic neurotransmitters, such as catecholamines and serotonin, are released by pathways descending from the locus ceruleus and hypothalamus, and nucleus raphe complex, respectively. Disruption of these descending inputs abolishes an important regulatory influence upon the intrinsic spinal circuitries. Direct intraspinal transplantation of the embryonic brainstem regions containing appropriate monoaminergic neurons was used to restore release of the missing neurotransmitters. This approach resembles in principle the prototype of a modern neurotransplantation experiment in the models of Parkinson's disease, in which grafts of embryonic substantia nigra were shown to restore dopamine release in the deafferented striatum, and is reviewed in Chapter 3.

In another series of experiments selected neuronal populations in the host spinal cord were replaced by homologous neurons derived from embryonic grafts. So far this has been attempted only for motoneurons of the anterior horn. This subject will be reviewed separately in Chapter 6. Finally, grafts of embryonic spinal cord have been used both experimentally and clinically to obliterate the formation of cysts (syringomyelia) and the consequences of this condition.[6-9] Indeed, this is the only instance where transplant of embryonic spinal cord into adult spinal cord has been used clinically.

Survival and Development of Embryonic Grafts in the Spinal Cord

Some early studies exploring various methods of spinal cord repair reported poor survival and retention of the grafts in the lesion cavities.[10,11] The subject was revisited in the 1970s and early 1980s with better results, but the spinal cord was still considered as being a difficult site for transplantation.[12-18] The increasing understanding of the biology of embryonic grafts, and improvements of surgical protocols were gradually adapted to the specific needs of the spinal cord. Even with this advantage the research on transplants in the spinal cord was lagging somewhat behind the dynamically expanding field of neurotransplantation in the brain, and it took a few years before the optimal conditions for grafting became established. Although the spinal cord appears to be relatively accessible, its small size in most experimental animals, particularly rats, a compact organization of its anatomical structures and its rich vascularization make it a challenging site for surgery. In the majority of studies the grafts of embryonic neural tissues were placed in an acute lesion cavity made by hemisection, bilateral posterior funiculotomy or "overhemisection" involving both posterior columns and the lateral/anterior columns on one side. In other experimental models a subpial injection of a fragment of embryonic tissue, or a volume of cell suspension, into the primarily intact spinal cord was used. The grafts generally do not survive and integrate well into the large cavities produced by complete transverse section.[18,19] One important prerequisite of survival and development of the grafts is their approximation to the rich vascular surface of the host allowing their prompt vascularization.[1,17-21]

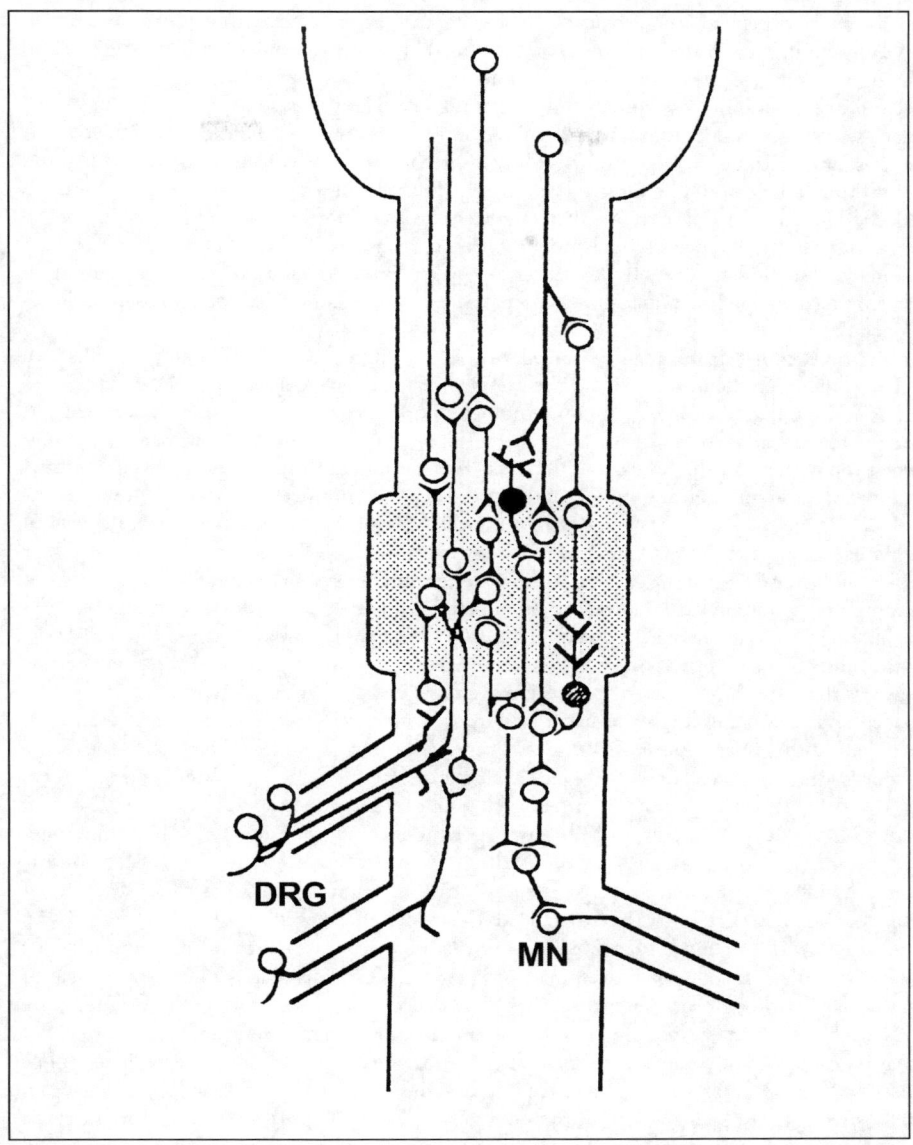

Figure 1. A diagram illustrating various ways in which an intraspinal embryonic graft (shaded area), introduced at the site of injury, may establish a novel spino-spinal relay for conduction of motor and sensory information caudally to the lesion. The top of the diagram represents cortical and brain stem regions. The various configurations of neural circuitries and the relative length of axonal projections into and out of the graft are based on experimental evidence discussed in the text. MN, motoneurone; DRG, dorsal root ganglia; note that the possibility of dendritic sprouting of the graft (solid circle) and host (hatched circle) neurons is also indicated. Reproduced with permission from ref. 76 (Reier et al. Prog Brain Res 1988; 78:173-179).

This event may depend on one hand on the function of the graft facilitating growth of capillaries from the adjacent host tissues, and, on the other hand, on the cellular responses triggered by the injury to the host spinal cord. It has to be noted that the model of acute total or subtotal section or laceration of the spinal cord is relevant to a relatively rare clinical situation in humans which may result from a missile injury or a penetrating stab wound. Nevertheless few studies have addressed the issue of feasibility of grafting into the contusion lesions or chronic injury sites which would be more relevant to the human trauma[22-24] (see also Bregman et al, 2002).[5] The role of trophic factors that enhanced the success of survival and integration of the graft into the adult spinal cord has recently been emphasized,[5] as well as that of intracellular messengers such as cyclic AMP.[25] Needless to say, the reproducibility of the lesion cannot be overemphasized when the impact of any repair procedure, including transplantation, on functional recovery is to be tested.[26]

An additional problem to be considered here is the degree of immunological tolerance achieved between the host CNS and the graft from an allogeneic donor. The CNS has always enjoyed a reputation of being an "immunologically privileged site" and this remains true, at least for practical purposes in the animal experiment. In the standard experimental situation when the donor and the host come from the same strain, and usually from the same inbred colony, the immunological rejection of the graft is not a major determinant of its survival.

Although there is some degree of immune response of the host, transplants in the central nervous system fare better than in other sites (for review see Freed et al, 1988).[27] Another powerful factor determining survival of embryonic neural grafts is the use of donor embryos of appropriate gestational age. The developmental time window during which optimal results can be obtained varies for each neuronal population, and grossly corresponds to the end stage/early postmitotic phase in formation of a given neuronal pool, before a significant extension of axons towards the target has occurred.[1,2] At this stage the mitotic potential of grafted tissue is maintained and direct axotomy of immature neurons, which is known to cause their death,[28] is avoided. Autoradiographic studies of development of the rat spinal cord have shown that the majority of neurons originate between days 11 and 16 of gestation (embryonic days 11-16; ED11-16) and that the sequence of neuronal pool proliferation follows both a rostral-caudal and a ventral-dorsal gradient.[29,30] The poor survival of spinal cord grafts obtained from older fetuses (ED16-18) reported by earlier studies[17,21] probably reflected the exhausted mitotic potential of the tissue and axotomy-induced death of many neurons. In contrast, grafts of embryonic spinal cord obtained from ED11-ED15 donors showed excellent long term (1-16 months) survival in more than 80% of animals, comparable to survival of grafts into the brain.[31-34] Survival and growth of spinal cord grafts in the neonatal and adult hosts appear to be similar,[34] although some earlier reports suggested that it could be better in the immature CNS.[35] However, the response of developing spinal cord to transplantation, and the degree of anatomical integration between the graft and the host are different than in the adult, as will be discussed below. Embryonic spinal tissue inserted into the lesion cavity grows to span the gap, and usually at least some regions of the severed spinal cord become reconnected by the graft. The degree of fusion on the graft-host interface is variable even within the same experimental animal, but there are usually some areas of confluent transition between the two neuropils (Fig. 2).

It has been demonstrated that the volume of grafts increases substantially during the first weeks after grafting as a result of proliferation of cells, both neurons and glia, and expansion of developing neuropil. In one study the increase in volume of ED13 spinal grafts injected into the primarily intact adult spinal cord was estimated as sevenfold.[31] Light and electron microscopic studies have shown that the homotopic grafts of embryonic spinal cord achieve maturity about 1 month after transplantation, after which time their morphology does not change significantly.[33,34] One important finding was that even after survival periods as long as 16 to 24

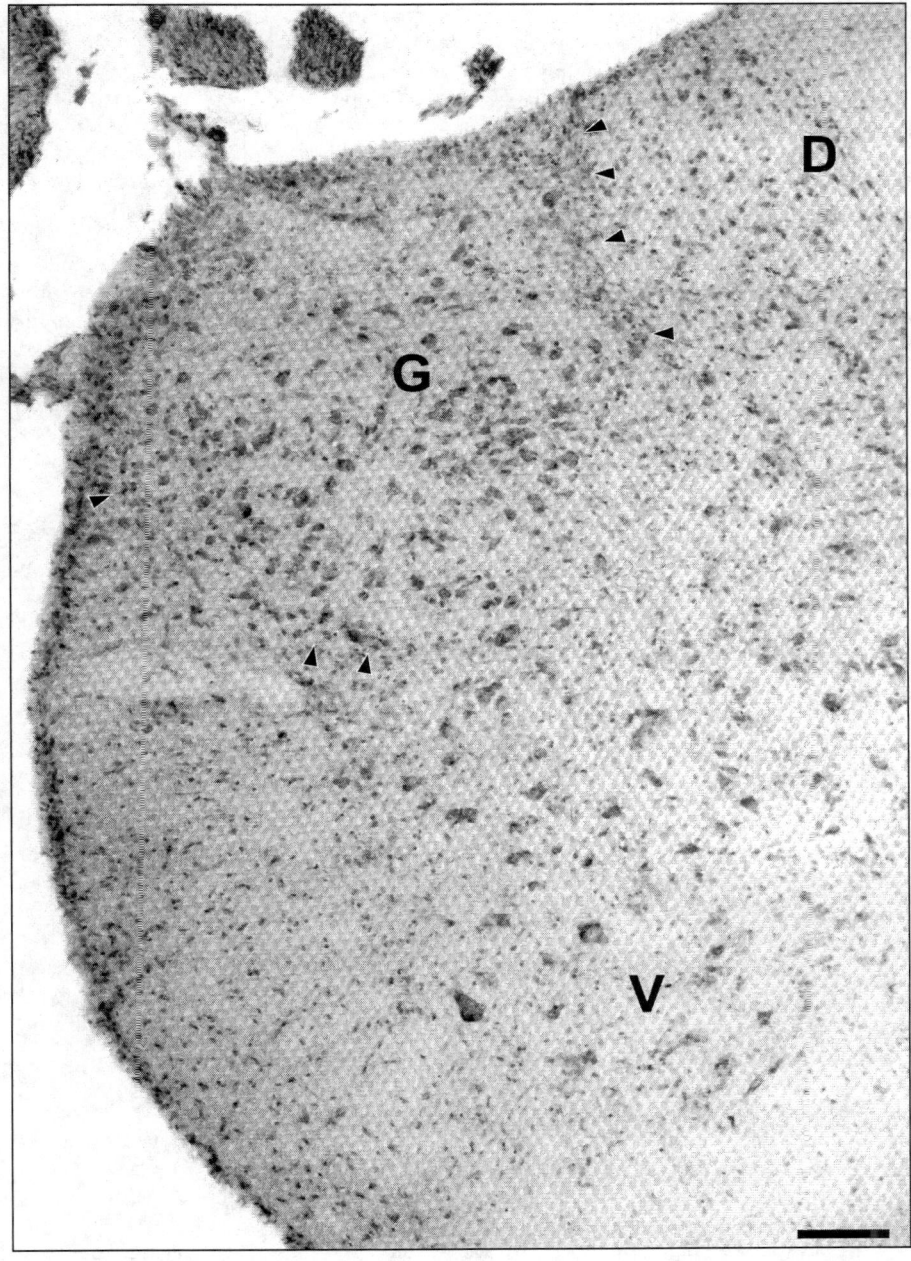

Figure 2. Transverse cryostat section stained with cresyl violet showing the position of the graft within the L6 segment of the spinal cord. The dorsal horn on the operated side is marked D, the ventral horn V. The arrowheads show regions of moderate gliosis, but there are also areas of good fusion of the graft with the host. Scale bar = 100 μm. Reproduced with permission from ref. 42 (Clowry, Vrbová. Exp Brain Res 1992; 91:249-258).

months[34] (Sieradzan, personal unpublished observations) the grafts do not exhibit any obvious signs of necrosis. Thus, at least in the rat, the graft can remain viable for the life time of the animal. The only pathology consistently seen in the transplants is hypertrophy and hyperplasia of astrocytes as revealed by increased glial fibrillary acidic protein (GFAP) immunoreactivity. This is particularly striking in poorly integrated grafts which developed in apparent isolation from the host.[34] At the ultrastructural level the astrocytic processes within the grafts appeared thicker, more densely packed and had an increased intermediate filament content. Similar above-normal GFAP reactivity was seen in embryonic spinal cord grafts developing in the anterior chamber of the eye,[36] and in the grafts of fetal neocortex and hippocampus, both in oculo and in the brain.[37-39] This glial reaction may be a general feature of embryonic transplants. One possible explanation is that it represents a persistent response to the initial surgical trauma. However, the intensity of the glial reaction increases with time after transplantation. For example, in ED14 neocortical and cerebellar transplants into the cerebral cortex GFAP immunoreactivity was first detected one week postoperatively and reached, or even surpassed that seen in the adjacent tissues of the host after 30 days in situ.[40] The time course of these events suggests that the glial reaction could reflect disturbed cell interactions in a graft developing in isolation from its physiological afferent inputs inducing the glial processes to fill the unoccupied synaptic sites.[34]

Mature grafts of embryonic spinal cord have a relatively reproducible morphology regardless of the site of transplantation. The overall appearance of the grafts developing in oculo,[36,41] in the adult cerebral cortex[32] or in the spinal cord of neonatal or adult rats was similar with regard to neuronal composition, degree of myelination and organotypic differentiation.[34,42,43] One to 16 months after transplantation the ED14 spinal cord grafts contained a mixture of evenly spaced, well developed neurons predominantly of small and medium size. In light microscopy a variety of neuronal shapes could be seen. The majority of cells had a mature bipolar or multipolar appearance, but even as long as 6 weeks postoperatively some neurons displayed features of immaturity such as rounded shape, a high nuclear/cytoplasm ratio and sometimes double nucleoli. At the ultrastructural level the neurons in the graft had well developed stacks of rough cytoplasmic reticulum, prominent Golgi complexes and abundant mitochondria.[33,34] Morphometric analysis of their sizes revealed a similar range of soma diameter (12.5-40.29 µm), but with a different distribution than in the intact adult spinal cord. One feature of ED14 spinal grafts was a relative paucity of larger cell bodies of the size comparable to alpha motoneurons. Because motoneurons are the first population emerging in the embryonic spinal cord, and by ED14 many will have already extended their axons to the peripheral targets,[44] the paucity of motoneurone-like cells in this preparation could be explained by their selective death due to direct axotomy and/or to the deprivation of target at the critical developmental stage.[45,46] Consistent with this possibility was the finding that spinal grafts harvested from ED12 embryos were enriched in large cell bodies resembling motoneurons.[32]

Although grafts of embryonic spinal cord do not develop into "mini spinal cords", detailed anatomical studies have revealed that, similarly to grafts of some structures of the embryonic brain,[47-54] they achieve some degree of organotypic differentiation characteristic of the mature tissue. Thus, Golgi studies of ED14-15 spinal grafts developing in oculo revealed that neurons with similar morphological features were clustered together, reproducing to some extent organization into the ventral horn-, intermediate grey- and dorsal horn-like areas.[55] Another feature observed in ED14 spinal grafts was the presence of myelin-free regions populated by small neurons (7-15 µm in diameter), many of which immunostained with antibodies against neuropeptides like substance P, met- and leu-enkephalin, neurotensin and somatostatin.[56] These myelin-free zones closely resembled substantia gelatinosa in the superficial dorsal horn of the adult spinal cord. The similarities were striking, but some important differences in the distribution of peptidergic neurons and organization of the neuropil were also detected. For example, unlike in the adult spinal cord, some neurotensin- and enkephalin-positive neurons were also found outside the myelin-free zones. Moreover, the scalloped terminals and related glomeruli characteristic of normal primary afferent innervation of the substantia gelatinosa were not found in the myelin-free regions of the grafts[56] despite the fact that the primary afferent fibres are known to regenerate into the juxtaposed embryonic spinal cord.[57] One important question raised by these and similar experiments is to what extent the observed cytological and organotypic differentiation of the spinal grafts reflects their intrinsic potential, and to what extent it is modulated by the afferent and efferent connections with the host CNS. Findings that some degree of differentiation is seen in grafts of embryonic spinal cord developing in oculo[41,55,58] and in organotypic explants in vitro[59,60] argue against this possibility. In the experiments of Jakeman and colleagues differentiation of superficial dorsal horn-like regions was similar in the grafts developing in the spinal cord or in the cerebral cortex, where no interaction with the physiological afferents could occur. However, the aberrant morphological features frequently seen in the grafts might reflect the lack of fine developmental "tuning" normally provided by appropriate afferent innervation.

In summary, with improvements in surgical technique and optimal selection of the gestational age of donor tissue, homotypic embryonic grafts provide a reproducible experimental model for studies on anatomical and functional recovery of the host spinal cord. A reasonable degree of differentiation achieved by the grafts should be an asset if appropriate reconnection of the host circuitry is to be expected, as might be suggested by highly specific graft-host interactions seen in some brain models.[61-67]

Integration of the Grafted Tissue with the Host Spinal Cord

Development of the Graft-Host Interface

The anatomical integration between the graft and the host requires reciprocal axonal connections which can only be established if there are no major barriers impeding axonal growth. The CNS responds to injury by formation of a dense glial scar consisting of a newly established glia limitans (accessoria), basal membrane and fibroblasts which was thought to be a mechanical barrier for regenerating axons.[68-72] This process occurs within 2 weeks after injury. Therefore axonal projections between the graft and the host have to develop either very early before the intervening gliotic scar is present, or can develop later in the areas where the formation of the scar has been prevented. Some experimental data show that embryonic grafts can modify the natural history of cicatrisation in the CNS.[32,73] It is instructive to compare the sequence of early events that follow a lesion inflicted to the CNS with no graft in place, and in the presence of acutely transplanted embryonic neural tissue. Large suction cavities were made in the occipital cortex and hippocampus overlying the superior colliculus. Grafts of embryonic cerebral cortex and cerebellum were used.[73] During the first postoperative week the zone of necrosis around the cavity became invaded by macrophages removing the tissue debris and the lumen of the cavity filled with fibroblasts of meningeal origin (see also Krikorian, 1981).[74] After one week most of the debris in the grey matter had been removed and the cavity was surrounded by an incomplete layer of strongly GFAP-positive astrocytes with their processes aligned towards the lumen. In the white matter cysts surrounded by astrocytes formed. At 10 days the scar appeared mature and remained unchanged up to 60 days postoperatively. A new glia limitans consisting of one to several layers of astroglial processes, covered with laminin-positive basal lamina and meningeal cells, developed in the regions of the grey matter, while in the white matter cystic cavitations progressed even further.

Grafts placed into the lesion cavity altered the processes described above. After the first week no intervening glia limitans or basal membrane could be identified in the sites of initial close apposition between the graft and the host tissue. Although the host astrocytes showed signs of activation, they did not line up at the interface with the graft. At 10 days postoperatively, in most regions a complete fusion with no astrocytic border or basal lamina was established and remained there permanently. During the same time grafted tissue grew filling more or less completely the lesion cavity. However, in these later stages the graft made contact with the host tissue in which the process of formation of glia limitans accessoria was already advanced resulting in less complete fusion. If the expanding graft contacted the wall of the lesion cavity before the glia limitans was fully established (i.e., before 10 days postoperatively) some degree of fusion was usually achieved. However, if this approximation occurred later, both the graft and the host elaborated separate GFAP- and laminin - positive glia limitans with invasion of fibroblasts, and this practically precluded any axonal exchange between them. Thus, three distinct types of interface could be distinguished from complete fusion to extensive scar formation, depending on the degree of initial apposition of the graft to the acutely injured host tissue. Interestingly, some lines of evidence suggest that embryonic grafts could also induce a partial regression of an established gliotic scar surrounding a chronic hemisection cavity in the spinal cord of adult rats.[75,76] When ED14 spinal cord grafts were placed in such cavities some zones of uninterrupted fusion between the graft and the host were detected. Areas devoid of GFAP immunoreactivity represented on average 15 to 35% of the total area of graft-host interface although there was considerable variation in the outcome between individual experiments. Obviously, in this type of experiment it was difficult to control for disruption of gliotic scar in the host during transplantation but the results are, nevertheless, intriguing. What are the possible mechanisms by which embryonic grafts interfere with formation of gliotic scar? Simple approximation of the injured neuronal surfaces is not a sufficient explanation because, in the adult CNS, both sides of a cut wound develop glial-meningeal scar even if their edges reestablish contact.[77] On the other hand equivalent lesions to the developing CNS do not result in scarring and are functionally less detrimental than in the adult.[78-81] Within the neural transplantation setting many lines of evidence point towards a role for immature astrocytes present in the grafts. As discussed in detail in Chapter 2 the role and behaviour of astrocytic glia during axonal regeneration are developmentally regulated; the immature astrocytes support axonal growth[82,83] while interaction of axons with adult astrocytes induces cessation of their growth.[84] There seems to be a critical period, occurring in the rat around 8 days postnatally, beyond which the properties of astrocytes switch from growth promoting to growth impeding.[85] It is possible that during de novo formation of the graft-host interface the embryonic tissue releases factors inhibiting glial proliferation and preventing alignment of the astrocytic processes in the host. Similarly, graft could limit invasion of mesenchymal cells by antagonising activity of fibroblast growth factor (FGF) and akin molecules in the lesion site, or by releasing enzymes degrading the components of basal lamina, such as plasminogen activators.[86] These local effects could be mediated, at least in part, by immature astrocytes present in the graft, some of which can migrate into the host tissue.[87]

Evidence for Axonal Projections between the Graft and the Host

The rationale for using embryonic grafts to bridge the lesion site in the severed spinal cord is based on two assumptions: (1) the grafts could act as a substratum conducive to axonal regeneration and/or as a source of neurite growth promoting trophic factors; (2) the grafts could provide a source of neurons to replace the lost grey matter of the host and establish a novel spino-spinal relay network (see Fig. 1). The desirable final outcome would be anatomical and functional reconnection of the spinal cord. This might occur through the formation of mono- or polysynaptic circuitry involving ingrowing host axons, grafted neurons and their intrinsic connections, and axons from the graft projecting to the host.[24,88] The most important goal would be reinnervation of the motoneurone pools caudal to the lesion which could permit some degree of voluntary motor control. The patterns of axonal projections developing between the grafts of embryonic spinal cord and the host have been extensively studied using a variety of anatomical techniques.

Fewer studies have examined the graft-induced functional recovery. We will review the anatomical data from experiments carried out on (i) adult and (ii) developing animals and then (iii) move towards the functional implications of grafting.

Adult Animals

The degree of connectivity between the graft and the host can be examined using axonal retro- and anterograde tracing methods. Small injections of HRP/WGA-HRP mixture local-ized to the grafts resulted in retrograde labelling of numerous neurons well beyond the area of any detectable spread of the tracer. This suggested the existence of an extensive intrinsic con-nectivity within the graft.[24,34,89] Additionally, some anterogradely filled axons of the graft neu-rons were seen crossing the interface and could be traced in the dorsal and ventral grey matter. Injections of HRP into the host's parenchyma at distances of 5-7 mm from graft-host interface resulted in retrograde labelling of neurons within the graft. However, no labelling in the graft was observed if the tracer was injected at greater distances (Fig. 3). Based on the combined evidence from antero- and retrograde tracing studies, it has been estimated that these axons can elongate for up to 4-5 mm in the host neuropil.[34,89] Similar distances of axonal outgrowth have been reported for embryonic locus coeruleus neurons grafted into the adult spinal cord.[19] The labelled axons originating in the graft terminated near the cell bodies of the host and occasion-ally could be seen in the immediate vicinity of the motoneurons.[89] However, the ultrastruc-tural evidence of synapse formation with the host neurons is still lacking. The crossing axons were typically found in areas of confluent fusion with the host neuropil. In the zones where a gliotic partition was present, the labelled axons were oriented parallel to the interface with no evidence of traversing to the host. The number of graft neurons projecting to the host grey matter was highly variable. These neurons were usually clustered within 1 mm of the graft-host interface, with progressively fewer cells located towards the centre and the opposite pole of the graft[24] (see also Fig. 3).

The reciprocal host to graft axonal projections followed a similar pattern. Application of the tracers into the host parenchyma labelled a dense network of axons in the grey and white matter, but this network terminated quite abruptly at the interface with the graft. Only a few axons, sometimes in small bundles, crossed into the graft and these terminated within 0.5-1 mm. The predominantly local character of the host to graft projections was further confirmed by retrograde labelling with the tracers injected into the graft. Small central injections yielded only a few retrogradely labelled neurons in the adjacent intermediate grey matter. The retro-grade tracers injected closer to the interface labelled a moderate number of small to medium sized neurons in the medial and lateral intermediate grey matter scattered within 0.5-1 mm of the graft-host interface. In summary, these projections were limited, occurring at best at the segmental level, and sometimes involving also the ipsilateral dorsal root ganglia[89] (see also Tessler et al, 1988).[57] The absence of retrogradely labelled neurons in more rostral segments of the spinal cord, in the brain stem and in the cerebral cortex was somewhat disappointing. Could these limited and short distance projections be of any functional value?

There is also some evidence of limited ingrowth of long descending axons into the trans-plants of embryonic spinal cord. For example immunocytochemical studies revealed the pres-ence of the serotonergic fibres within the grafts.[23,34] These fibres represented either regenera-tive growth, or sprouting of the injured host axons descending from the brain stem. Similarly, the corticospinal fibres were found to regenerate/sprout into the grafts placed in the hemisec-tion cavity.[90] However, in both cases no evidence of reinnervation of the host's spinal cord caudal to the transplant was obtained.

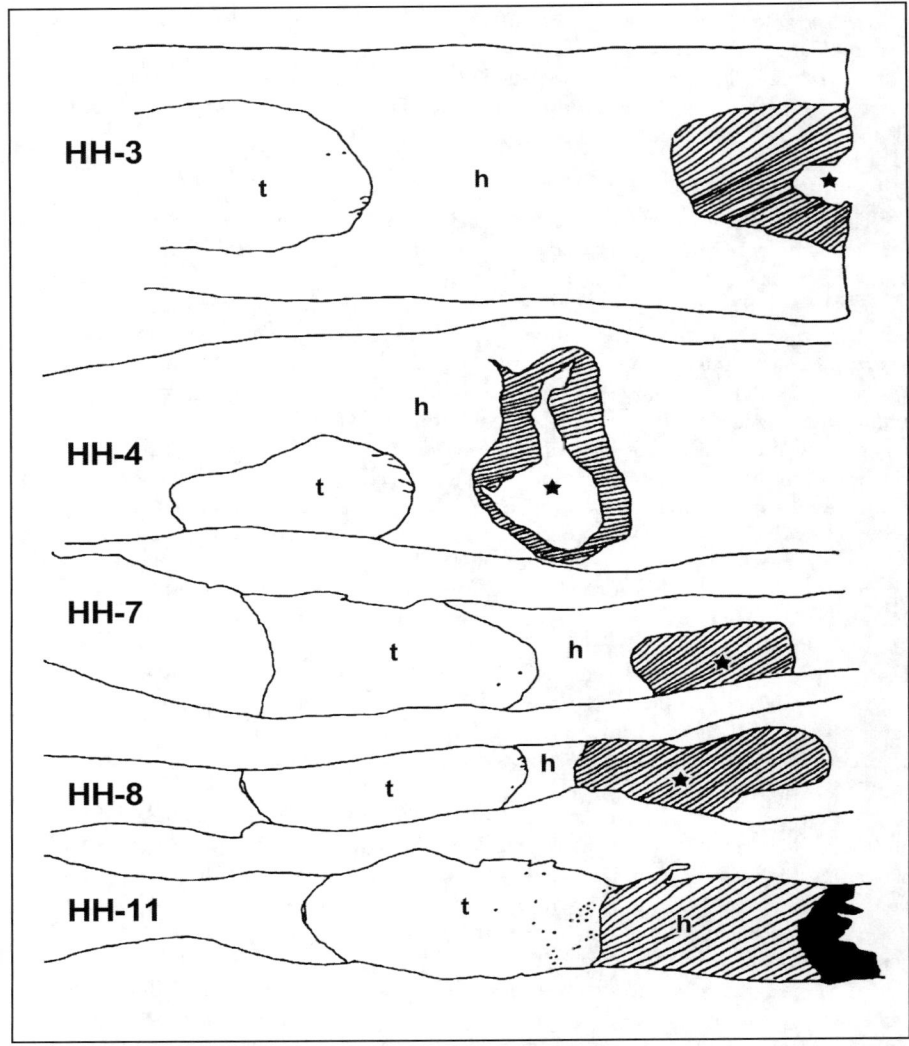

Figure 3. Axonal projections from the embryonic spinal cord graft into the adult rat spinal cord. Tracings of sections from five transplants (t) with HRP/WGA-HRP injections (shaded areas) into the host tissue made at varying distances from the graft-host interface. (*) indicates damage to tissue at the injection site. Few retrogradely labelled neurons, clustered close to the interface, were found in grafts in four out of five animals; each dot represents one labelled neurone. Anterogradely labelled axons, crossing from the host to the graft, were present in three recipients. Scale bar = 1 mm. Reproduced with permission from ref. 89 (Jakeman et al. J Comp Neurol 1991; 307:311-334).

In summary, the anatomical data show the existence of the reciprocal, mostly local connections between the graft and the host. However, the numbers of neurons involved in their formation, particularly on the host's side, seem to be quite limited, and the distances which can be negotiated by the axons growing from the graft are short. Although it is possible that the present techniques tend to underestimate the extent of connectivity achieved, the graft-induced regenerative capacity of the adult spinal cord still appears to be at best moderate. The situation is different in the spinal cord of immature animals.

Developing Animals

Most of the experimental data on the graft-induced regeneration and plasticity of the immature spinal cord were obtained in the rat. As discussed in detail in Chapter 3, at birth the development of the rat spinal cord is far from complete. Although the dorsal root afferents and most of the inputs descending from the brain stem are already in place other descending fibres develop mostly postnatally. The corticospinal pathways, for example, do not reach the lumbar region before postnatal day 7 (PD7), and some late growing axons are still added to the established pathways during the first days of life. These developmental characteristics allowed the design of some interesting experimental models.[91,92]

Unlike in the adult spinal cord, where after injury to the long descending tracts no significant regeneration occurs, in the neonatal animals both late growing and regenerating axons are able to circumvent the lesion site. Such axons take an aberrant route through the undamaged adjacent grey matter and reach their normal targets caudal to the lesion[93,94] (see Chapter 2 for discussion). Plasticity of the system decreases as the animal matures, and finally becomes exhausted during the third postnatal week. This unique response provides an anatomical basis for the functional recovery observed after neonatal lesions.[94-99] After transplantation of embryonic spinal tissue into the injured neonatal spinal cord short distance axonal projections between the graft and the host are formed at the segmental level. The pattern of connectivity is similar to that observed in the adult but its extent is much greater. In particular, axons derived from the graft can elongate for much longer distances in the host neuropil. For example injections of HRP/WGA-HRP as far as 20 mm from the graft-host interface still revealed some retrogradely labelled neurons in the graft.[34] Two other graft-related phenomena unique to the injured neonatal spinal cord occur. First, the grafts permit axonal growth through the lesion site into its caudal part. This has never been seen in adult animals in which the long axons penetrate into the grafts but are unable to traverse the full extent of the lesion even with optimal approximation of the graft and host tissues. One convenient experimental approach to this issue is looking at the descending serotonergic projections from the raphe nuclei in the model of the neonatal hemisected cord. These projections are exclusively supraspinal (except for a few local serotonergic cell bodies which appear in the spinal cord early in the animal's life) and can be easily identified by immunocytochemistry.[91] After hemisection serotonergic immunoreactivity in the grey matter below the lesion is dramatically reduced, although not completely abolished due to both the ability of some ipsilateral axons to circumvent the gap, and to the sprouting of the intact fibres crossing from the contralateral hemicord.[91] With the ED14 spinal cord graft in the lesion cavity the immunoreactivity of the host grey matter caudal to the hemisection was restored to more than 80% of normal.[91] The grafts also received a dense serotonergic innervation.[100] These results could be explained by various graft-induced events. First, the graft could act by preventing death of the axotomized projection neurons[101] which is known to happen after a neonatal spinal cord lesion (see Chapter 2). Secondly, the graft could enhance the aberrant regenerative growth and sprouting of the host fibres. Thirdly, the graft could act as a true tissue bridge permitting axonal growth breaching the entire lesion site. This latter response, unique to the neonatal spinal cord, was confirmed by subsequent experiments which showed that, indeed, axonal growth occurred across the transplant.[101] One important issue was whether the axons traversing the graft were truly regenerating, or whether they were late-arriving fibres added to the long tracts during the first postnatal days. This question was addressed in double labelling studies in which, before a graft was placed into the mid-thoracic hemisection cavity, neurons projecting to that level from the brain were prelabelled with a retrograde fluorescent tracer. Several weeks after transplantation another retrograde tracer was injected to the host spinal cord caudal to the graft.[92] Examination of the brains of the host animals revealed that a variety of subcortical nuclei, known to project to the spinal cord, contained both double-labelled cells and cells single-labelled with the second tracer.

Thus, both regenerating and late-arriving axons had entered the spinal cord caudal to the graft. The patterns of labelling seen in the individual subcortical structures reflected their developmental history (see Chapter 2). For example, neurons of the red nucleus contained either the first tracer or were double-labelled, indicating that their axons predominantly regenerated into, and beyond, the graft. The pattern of labelling in the locus coeruleus and in the raphe nuclei was consistent with both regeneration and late growth of the axons across the graft. In contrast, the neurons of the sensorimotor cortex, whose late-arriving axons were not directly injured by the surgery, uniformly displayed labelling only with the second tracer. The late-developing corticospinal tract is a convenient system for testing how the developmental plasticity can be induced by embryonic grafts[90] (see also Chapter 2). A bilateral posterior funiculotomy within the first two days of the animal's life disrupts the future trajectories along which the tract should develop. As mentioned above, after such a lesion some of the arriving axons become rerouted into the adjacent grey matter and circumvent the lesion. When embryonic spinal cord was placed in the lesion cavity, in addition to the usual aberrant growth in the intact grey matter, the corticospinal axons formed a dense plexus throughout the entire graft and many of them crossed to the host caudal to the lesion. The final reinnervation of the spinal cord below the lesion was much greater than in the animals without grafts. A similar lesion inflicted between PD5 and PD8, i.e., after the nerve fibres have reached the level of the lesion, but prior to synaptogenesis, resulted in axotomy of the corticospinal fibres. Nevertheless, a modest growth around the lesion site occurred in about one third of the animals, indicating that some axons still retained their regenerative capacity. Again, embryonic grafts consistently improved reinnervation of the host spinal cord below the lesion. Later lesions, inflicted when the corticospinal fibres were making synaptic contacts in the spinal grey matter, i.e., from PD16 onwards, resulted in the adult-type response with no regenerative growth around the lesion and retraction of the axons. However, embryonic spinal cord placed in the lesion cavity prevented retraction of the corticospinal axons, which instead grew into, and terminated within, the graft close to the interface. Thus, the presence of the graft in the lesion site enhanced and prolonged the period of anatomical plasticity, which to a limited extent naturally occurs in the developing spinal cord after injury. It appears that gradual exhaustion of this plasticity coincided with target finding and synaptogenesis in the grey matter.

A number of factors, both intrinsic to the severed axons themselves and to their environment, could determine the extent of regeneration and account for the differences between the neonatal and adult experiments (see also Chapter 2). In both experimental situations embryonic grafts present the same set of opportunities for the host axons, but there appears to be a critical period beyond which the anatomical plasticity of the system becomes greatly restricted. This transformation correlates with maturation of the glia in the host spinal cord,[102] which, as discussed above, switch from the growth supporting to growth inhibiting mode.[85] Accordingly, it is possible that maturation of astroglia in older animals could hamper elongation of axons into, and out of, the grafts, and limit the aberrant axonal growth around the lesion cavities. Another factor could be the developmentally regulated expression of surface molecules in the host spinal cord, which might either promote or inhibit axonal growth. For example, with development of the host oligodendroglia and deposition of myelin, growth inhibitory myelin-associated proteins would progressively limit elongation of axons.[103,104] In contrast, immaturity of astroglia and the initial absence of myelin in the transplants would create a favourable milieu for axons of the host. Additionally, neural cell adhesion molecules (N-CAM) and the components of extracellular matrix such as laminin, known to support axonal elongation in vitro and in vivo, could also facilitate regeneration in the developing spinal cord.[105-108] The presence of laminin, fibronectin and heparin sulfate proteglycan has, indeed, been demonstrated in embryonic, but not in mature, spinal cord.[106,107] However, the role of these molecules is still not clear; for example the trajectories of the corticospinal pathways do not express laminin during their postnatal development.[109] Trophic support lent by the grafts to the injured host neurons has been discussed in detail in Chapter 2. Briefly, after a neonatal spinal cord lesion the axotomized supraspinal projection neurons can be permanently rescued from cell death only by homotopic grafts of embryonic spinal cord. The neuroprotective effect of heterotopic grafts is only temporary and the cells finally die.[110] These observations suggest that the trophic support, which can be provided by a variety of immature neural tissues, is sufficient to prevent cell death acutely after injury, but long-term stabilization of the system relies on the formation of synaptic contacts with neurons within the graft, which serve as foster targets.

The regenerative response of host neurons may also depend on their developmental stage. For example, graft-induced developmental plasticity of the corticospinal axons decreases dramatically around the time when they switch from growing mode to target-finding and synaptogenesis.[90] Such axons do not invade the deeper parts of the grafts but terminate shortly after crossing the interface, similarly to axons in the adult spinal cord. It seems that after a critical developmental stage, probably corresponding to the onset of synaptic activity, the axons are programmed to rapidly find an alternative target among the grafted neurons. This cannot be overcome even by the growth-permitting signals from the environment. In contrast, axons which are still in the growing mode, continue growing in the favourable environment of the transplant and finally reach the spinal cord caudal to the graft (Fig. 4).

In summary, the degree of connectivity between the graft and the host is determined by complex interactions between the two partners. These seem to involve (i) the developmentally regulated potential of the host axons to grow and/or regenerate, (ii) the ability of host neuropil to support axonal elongation (also developmentally regulated), (iii) the trophic and neurite growth-promoting molecules produced by embryonic grafts, (iv) the favourable terrain for axonal elongation provided by the grafts and (v) the ability of the grafted neurons to serve as foster targets for the host axons.

Graft-Induced Improvement of Function

In the previous paragraph the anatomical basis for graft-induced recovery of locomotor function was discussed. The experimental evidence that such functional recovery indeed occurs is still limited compared to the amount of neuroanatomical data available. Perhaps one of the reasons for this discrepancy is that behavioural experiments tend to be more difficult than neuroanatomical studies. However, the interest in functional aspects of neurotransplantation is growing, and a few studies addressing this extremely important subject have recently been published.[111-115]

The survey of the anatomical data suggests that the basic apparatus, through which the functional recovery could occur, is restored to an appreciable degree, especially in the neonatal hosts. Thus, short distance reciprocal connections between the graft and the host are established at the segmental level. Unfortunately, these connections are relatively sparse, particularly in the adult animals. Also the primary afferents regenerate into the grafts but are unable to cross to the ventral horn to reestablish a monosynaptic reflex arch. Nevertheless, there is a possibility that some sort of polysynaptic connections with motoneurone pools, involving neurons in the graft, do exist. As far as the long descending pathways mediating the control of locomotion are concerned, their regeneration/sprouting in adult animals is very restricted, and they terminate soon after entering the transplant.

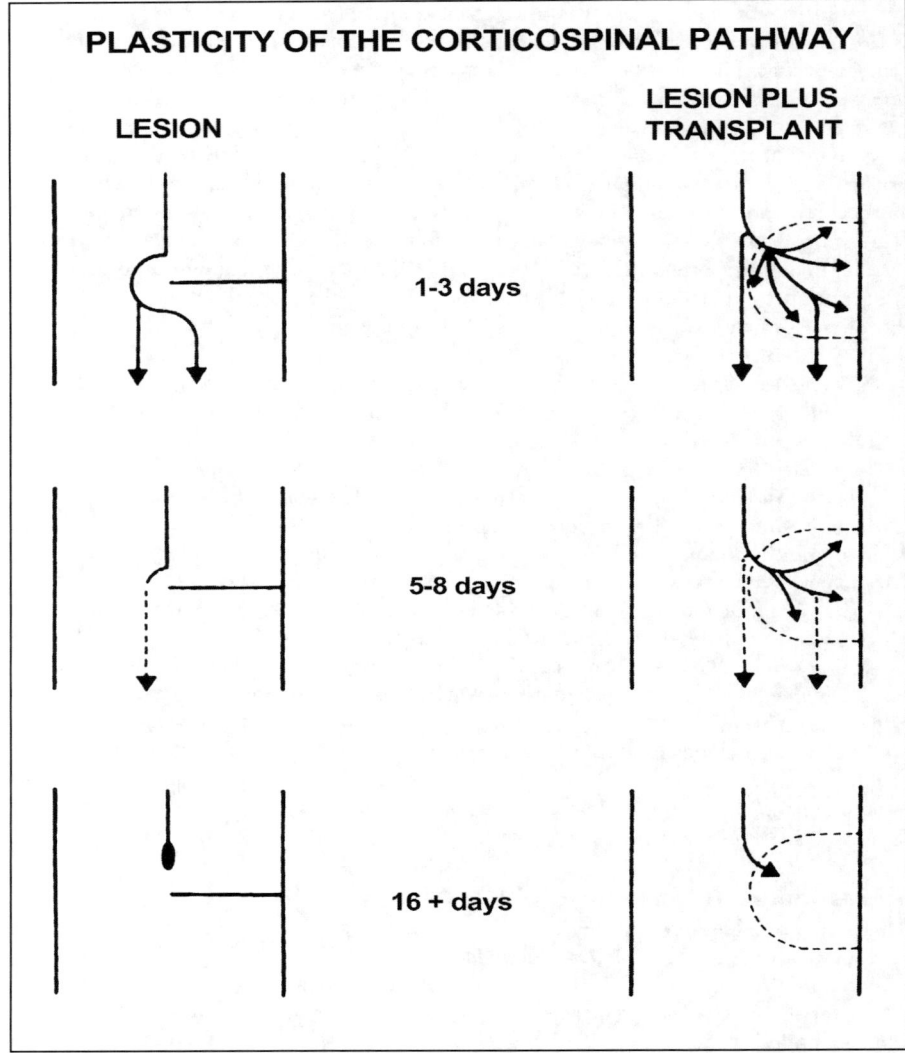

Figure 4. The diagram summarizing the enhancement of developmental plasticity of the injured corticospinal tract by grafts of embryonic spinal cord placed into the lesion cavity. See text for description. Reproduced with permission from ref. 90 (Bregman et al. J Comp Neurol 1989; 282:355-370).

The model of "infant lesion" to the spinal cord has been used to examine how the intervention of the transplant would modify the pattern of recovery of locomotion. In one study, carried out on neonatal rats, the spinal cord was "overhemisected" and embryonic spinal cord was grafted into the lesion cavity.[111] Eight to 12 weeks later the animals were trained to walk on a treadmill and to cross a grid runway in order to reach water reward. It is thought that overground locomotion, like crossing a runway, depends on the supraspinal motor control. In contrast, quadrupedal locomotion on a treadmill does not require descending control, but depends on the integrity of propriospinal connections between the forelimb and hindlimb segments. General qualitative observation and quantitative analysis of the footstep prints were used to assess the gait pattern of animals. The results showed that although both lesioned controls and graft recipients were capable of considerable locomotor activity, the latter group developed a less abnormal gait.

Typically, after a neonatal injury the animal's base of support during walking is broad, and the hindlimb contralateral to the lesion is abnormally rotated. Unlike the lesioned controls, the animals with grafts had a normal base of support, little contralateral limb rotation and, additionally, a compensatory rotation of the ipsilateral hindlimb. The animals with grafts also crossed the runway faster and with significantly fewer erroneous foot placements. This pattern of recovery involved bilateral changes in limb placing. These bilateral changes distinguish the outcome of a neonatal hemisection from that in the adult.[97] The pattern of locomotion is permanently reorganized so that motor control of the less affected hindlimb is adjusted to compensate for defective function of the more affected one.[97] This adjustment would ultimately result in more balanced function although apparently at the price of greater deficit on the less affected side. The beneficial effects of grafts on functional recovery could be explained by their ability to promote and extend the developmental plasticity of the host spinal cord.

Some experimental data are also available on recovery of locomotor function in kittens after complete transection of the spinal cord and transplantation of embryonic spinal cord into the lesion cavity.[113] In spinal kittens reflex bipedal walking on a treadmill can be readily elicited (and even develops precociously compared to intact animals), but the hindlimbs are not used spontaneously for overground locomotion. Interestingly, the spinal kittens can be trained to perform some quadrupedal treadmill walking, but this locomotion is very unstable with frequent falls and little weight-bearing on the hindlimbs. Spinal kittens which received embryonic grafts developed significantly improved pattern of locomotion although they never achieved a mature level of performance. In these animals quadrupedal locomotion appeared earlier than in the lesioned controls and their hindlimbs provided greater support with less frequent falls. In contrast with spinal controls, the recipients of transplants also developed overground locomotion although this was delayed by about two weeks compared to their intact littermates. Most importantly, there was a degree of coordination between the forelimbs and hindlimbs during the overground and treadmill locomotion, a feature which is never present in spinal kittens. These results, and in particular the presence of overground locomotion including forelimb-hindlimb coordination, suggest that some degree of reconnection of the spinal cord must have occurred. As discussed above, this could involve growth of the developing and/or regenerating axons across the transplant into the caudal portion of the spinal cord and formation of relay network within the graft. Moreover, the graft could act by preventing retrograde death of neurons within the spinal locomotion generators whose ascending axons might have been disrupted by spinal transection.[113]

Recent data obtained in adult rats and cats show that some aspects of functional impairment due to the injury to the mature spinal cord may also be amenable to grafting.[112,114] The model of spinal cord injury used in these studies was contusion produced at the lower thoracic level by weight-drop method, which, as often emphasized, is pathologically more relevant than transection to most of the cases of human spinal cord trauma. Ten days after spinal cord contusion in adult rats cell suspension grafts prepared from embryonic spinal cord were introduced into injury cavity. Both spontaneous (Tarlov open-field, inclined plane and grid walking) and conditioned (footprint analysis during runway crossing for food reward) bilateral hindlimb functions were studied for three months after injury. During this observation period the animals which received grafts did not show any improvement of spontaneous motor activity compared to the lesioned controls. However, detailed footsteps analysis revealed that certain features of gait, such as base of support and stride length, were improved by the presence of the transplants. These effects were observed in both hindlimbs. In contrast with the neonatal experiments described above, the angle of limb rotation during walking remained permanently abnormal. These findings suggest that even in the adult spinal cord some graft-induced functional improvement does occur. The mechanisms by which the grafts influenced the function of the adult host remain poorly understood. Contusion injury produces bilateral damage to some of the long pathways, while sparing the others. It is possible that the functional "take over" by the remaining pathways might contribute to the observed improvement of function. Even more likely, the grafts may be influencing the intraspinal circuitries of the host caudal to the lesion (see Stokes and Reier, 1992[114] for discussion). It has, indeed, been demonstrated that grafts of embryonic spinal cord neurons can modify the pattern of motoneurone excitability after contusion lesions to the adult spinal cord.[112]

This brief outline of the available data shows that the results obtained to date are encouraging, particularly in immature animals, although much further work is needed to explore fully the potential of grafts to promote functional recovery. For example, it would be important to understand what is the functional advantage of different patterns of graft-host connectivity observed in neuroanatomical studies. Another fascinating issue is the graft-induced functional plasticity of the spinal cord, especially in adult animals. These and other questions still await answer in the long journey towards solving the enigma of spinal cord repair.

The Use of Other Cell Types to Promote Recovery After Spinal Cord Lesion

The possibility that recovery of spinal cord function can be promoted by transplants of Schwann cells, glial cells, macrophages or stem cells that assume the phenotype of glial cells is discussed in Chapters 4 and 7. Here it is pertinent to mention briefly experiments where stem cells were transplanted into injured spinal cord.

Stem Cells

Stem cells are the current fashion for possible repair of various tissues after injury or disease. They are favoured because they appear to be pluripotential and can be obtained from the same individual where repair is needed so that there is no danger of rejection.

However, the pluripotentiality of adult stem cells has been questioned, since it was shown that some adult stem cells do not turn into unexpected lineages as previously thought, but fuse with existing cells and assume the phenotype of the recipient. While this ability of adult stem cells to fuse with existing cells may be useful in some cases, it points out the difficulties of the use of stem cells for repair.[116] In spite of this, several attempts at transplanting stem cells into injured spinal cord have been carried out and these showed that they differentiated predominantly into glial cells when grafted into the spinal cord.[117] Nevertheless, some beneficial effects of stem cell transplants into injured spinal cord have been reported.[118,119] However, in view of all the controversy surrounding this topic the mechanism by which stem cells grafted into an injured spinal cord may enhance recovery is hard to understand.

Concluding Remarks

Making concluding remarks on the subject of spinal cord repair using embryonic neural grafts is not an easy task. One can be either overoptimistic emphasizing the morphological evidence of regeneration, which otherwise would not have occurred, or too pessimistic, realizing that after all these years of laborious research the evidence for significant recovery of function is still limited. The neurotransplantation approach has undoubtedly extended our understanding of the regenerative capacity of the spinal cord and proved that embryonic neural grafts can replace the lost spinal neurons and/or restore the local neurotransmitter equilibrium. One wonders whether the limits of what can be achieved in terms of graft-induced recovery of function have been reached. Fortunately "no" seems to be the answer. There are numerous possibilities still to be explored. For example trophic factors, perhaps delivered directly to the vicinity of the lesion and graft, might encourage axonal growth and enhance the relay connections between the graft and the host. Maybe with present techniques regenerating host axons and reciprocal graft-host connections are too few to mediate a significant functional improvement. One can also speculate that cografting of embryonic spinal cord, as a relay tissue bridge at the lesion site, and embryonic monoaminergic neurons, as a source of regulatory neurotransmitters driving the novel spino-spinal network, could improve functional recovery. These and many other possibilities could, and hopefully will, be tested in the future.

References

1. Björklund A, Stenevi U. Intracerebral neural implants: Neuronal replacement and reconstruction of damaged circuitries. Ann Rev Neurosci 1984; 7:279-308.
2. Dunnett SB, Björklund A. Mechanisms of function of neural grafts in the adult mammalian brain. J Exp Biol 1987; 132:265-89.
3. Lindvall C, Björklund A. Prospects for neural transplantation in neurodegenerative disease. In: Kennard C, ed. Recent Advances in Clinical Neurology, Vol. 6. Edinburgh: Churchill - Livingstone, 1990:23-57.
4. Anderson DK, Howland DR, Reier PJ. Fetal neural grafts and repair of the injured spinal cord.Brain Pathol 1995; 5:451-7.
5. Bregman BS, Coumans JV, Dai HN et al. Transplants and neurotrophic factors increase regeneration and recovery of function after spinal cord injury. Prog Brain Res 2002; 137:257-73.
6. Falci S, Holtz A, Akesson E et al. Obliteration of a posttraumatic spinal cord cyst with solid human embryonic spinal cord grafts: First clinical attempt. J Neurotrauma 1997; 14:875-84.
7. Akesson E, Kjaeldgaard A, Seiger A. Human embryonic spinal cord grafts in adult rat spinal cord cavities: Survival, growth, and interactions with the host. Exp Neurol 1998; 149:262-76.
8. Akesson E, Holmberg L, Jonhagen ME et al. Solid human embryonic spinal cord xenografts in acute and chronic spinal cord cavities: A morphological and functional study. Exp Neurol 2001; 170:305-16.
9. Thompson FJ, Reier PJ, Uthman B et al. Neurophysiological assessment of the feasibility and safety of neural tissue transplantation in patients with syringomyelia. J Neurotrauma 2001; 18:931-45.
10. Sugar O, Gerard RW. Spinal cord regeneration in the rat. J Neurophysiol 1940; 3:1-19.
11. Brown JO, McCough JP. Abortive regeneration of the transected spinal cord. J Comp Neurol 1947; 87:131-7.
12. Nygren L-G, Olson L, Seiger A. Monoaminergic reinnervation of the transected spinal cord by homologous fetal brain grafts. Brain Res 1977; 129:227-35.
13. Thuline DN, Bunge RP. Preliminary observations on the transplantation of spinal cord tissue in rats. Anat Rec 1972; 172:418.
14. Olson MI, Bunge RP. Spinal cord transection: Results of implanting cultured embryonic spinal cord at the transection site. Soc Neurosci Abst 1974; 4:363.
15. Das GD. Neural transplants in the spinal cord of the adult rat. Anat Rec 1981a; 199:65A.
16. Das GD. Transplantation of neural tissues in the spinal cord of the adult rat. Soc Neurosci Abst 1981b; 7:625.
17. Das GD. Neural transplantation in the spinal cord of the adult mammal. In: Kao CC, Bunge RP, Reier PJ. eds. Spinal Cord Reconstruction. New York: Raven Press, 1983a: 3:67-96.
18. Das GD. Neural transplantation in the spinal cord of adult rats. Conditions, survival, cytology and connectivity of the transplants. J Neurol Sci 1983b; 62:191-210.
19. Nornes H, Björklund A, Stenevi U. Reinnervation of the denervated adult spinal cord of rats by intraspinal transplants of embryonic brain stem neurons. Cell Tissue Res 1983; 230:15-35.
20. Stenevi U, Björklund A, Svendgaard N-NA. Transplantation of central and peripheral monoamine neurons to the adult rat brain: techniques and conditions for survival. Brain Res 1976; 114:1-20.
21. Das GD. Neural transplantation in mammalian brain - some conceptual and technical considerations. In: Wallace RB, Das GD, eds. Neural Tissue Transplantation Research. New York, Heidelberg, Berlin: Springer Verlag, 1983c:1-64.

22. Winialski D, Houle JD, Jakeman LB et al. Transplantation of fetal rat spinal cord into longstanding contusion injuries of adult rat spinal cord. Soc Neurosci Abst 1987; 13:751.

23. Houle JD, Reier PJ. Transplantation of fetal spinal cord tissue into the chronically injured adult rat spinal cord. J Comp Neurol 1988; 269:535-47.

24. Reier PJ, Houle JD, Jakeman L et al. Transplantation of fetal spinal cord tissue into acute and chronic hemisection and contusion lesions of the adult rat spinal cord. Prog Brain Res 1988; 78:173-9.

25. Cai D, Qiu J, Cao Z et al. Neuronal cyclic AMP controls the developmental loss in ability of axons to regenerate. J Neurosci 2001; 21:4731-9.

26. Das GD. Perspectives in anatomy and physiology of paraplegia in experimental animals. Brain Res Bull 1989; 22:7-32.

27. Freed DJ, Drymechi J, Poltorak M et al. Intraventricular brain allografts and xenografts: Studies of survival and rejection with and without systemic sensitization. Prog Brain Res 1988; 78:233-41.

28. Lieberman AR. Some factors affecting retrograde neuronal responses to axonal lesions. In: Bellairs R, Gray EG, eds. Essays On The Nervous System. Oxford: Oxford University Press, 1974:77-105.

29. Nornes H, Das GD. Temporal pattern of neurogenesis in spinal cord of rat. Ist ed. An autoradiographic study - Time and sites of origin and migration and settling patterns of neuroblasts. Brain Res 1974; 73:121-38.

30. Altman J, Bayer SA. The development of the rat spinal cord. Adv Anat Embryol Cell Biol 1984; 85:1-166.

31. Patel U, Bernstein JJ. Growth, differentiation and viability of fetal cortical and spinal cord implants into adult rat spinal cord. J Neurosci Res 1983; 9:303-10.

32. Reier PJ, Perlow MJ, Guth L. Development of embryonic spinal cord transplants in the rat. Dev Brain Res 1983; 10:201-19.

33. Bernstein JJ, Patel U, Kelemen M et al. Ultrastructure of fetal spinal cord and cortex implants into adult rat spinal cord. J Neurosci Res 1984; 11:359-72.

34. Reier PJ, Bregman BS, Wujek JR. Intraspinal transplantation of embryonic spinal cord tissue in neonatal and adult rats. J Comp Neurol 1986; 247:275-96.

35. Hallas BH, Das GD, Das KG. Transplantation of brain tissue in the brain of the rat. IInd ed. Growth characteristics of neocortical transplants in hosts of different ages. Am J Anat 1980; 158:147-59.

36. Olson L, Björklund H, Hoffer BJ et al. Spinal cord grafts: An intraocular approach to enigmas of nerve growth regulation. Brain Res Bull 1982; 9:519-37.

37. Björklund H, Dahl D. Glial disturbances in isolated neocortex: Evidence from immunohistochemistry of intraocular grafts. Dev Neurosci 1982; 5:424-35.

38. Björklund H, Dahl D, Haglid K et al. Astrocytic development in fetal parietal cortex grafted to cerebral and cerebellar cortex of immature rats. Dev Brain Res 1983; 9:171-80.

39. Zimmer J, Sunde N. Neuropeptides and astroglia in intracerebral hippocampal transplants: An immunohistochemical study in the rat. J Comp Neurol 1984; 227:331-47.

40. Krüger S, Sievers J, Hansen C et al. Three morphologically distinct types of interface develop between adult host and fetal brain transplants: Implications for scar formation in the adult central nervous system. J Comp Neurol 1986; 249:103-16.

41. Henschen A, Hoffer B, Olson L. Spinal cord grafts in oculo: Survival, growth, histological organization and electrophysiological characteristics. Exp Brain Res 1985; 60:38-47.

42. Clowry GJ, Vrbová G. Observations on the development of transplanted embryonic ventral horn neurons grafted into the adult rat spinal cord and connected to skeletal muscle implants via a peripheral nerve. Exp Brain Res 1992; 91:249-58.

43. Sieradzan K, Vrbová G. Observations on the survival of grafted embryonic motoneurons in the spinal cord of developing rats. Exp Neurol 1993; 122:223-31.

44. Reynolds ML, Fitzgerald M, Benowitz LI. GAP-43 expression in developing cutaneous and muscle nerves in the rat hindlimb. Neuroscience 1991; 41:201-11.

45. Hamburger V. The Developmental History of the Motor Unit: The F.O. Schmidt Lecture in Neuroscience. Neuroscience Research Programme Bulletin. Cambridge, MA: M.I.T. Press, 1976; 1-37.

46. Lowrie MB, Vrbová G. Dependence of postnatal motoneurons on their targets: Review and hypothesis. Trends Neurosci 1992; 15:80-84.

47. Sotelo C, Alvarado-Mallart RM. Reconstruction of the defective cerebellar circuitry in the adult Purkinje cell degeneration mutant mice by Purkinje cell replacement through transplantation of solid embryonic implants. Neuroscience 1987; 20:1-22.

48. Kromer LF, Björklund A, Stenevi U. Intracephalic neural implants in the adult rat brain. Ist ed. Growth and mature organization of brain stem, cerebellar and hippocampal implants. J Comp Neurol 1983; 218:433-59.

49. Lund RD, Harvey AR. Transplantation of tectal tissue in rats. Ist ed. Organization of transplants and the pattern of distribution of host afferents within them. J Comp Neurol 1981; 201:191-209.

50. Mufson EJ, Labbe R, Stein DG. Morphological features of embryonic neocortex grafts in adult rats following frontal cortex ablation. Brain Res 1987; 401:162-7.

51. Isacson O, Brundin P, Kelly PAT et al. Functional neuronal replacement by grafted neurons in the ibotenic acid-lesioned striatum. Nature 1984; 311:458-60.

52. Isacson O, Brundin P, Gage FH et al. Neural grafting in a rat model of Huntington's disease: Progressive neurochemical changes after neostriatal ibotenate lesions and striatal tissue grafting. Neuroscience 1985; 16:799-817.

53. Isacson O, Dawbarn D, Brundin P et al. Neural grafting in a rat model of Huntington's disease: Striosomal-like organization of striatal grafts as revealed by acetylcholinesterase histochemistry, immunocytochemistry and receptor autoradiography. Neuroscience 1987; 22:481-98.

54. Clarke DJ, Dunnett SB, Isacson O et al. Striatal grafts in the ibotenic acid-lesioned neostriatum: Ultrastructural and immunohistochemical studies. Prog Brain Res 1988; 78:47-53.

55. Broton JG, Yezierski RP, Seiger A. Intraocular grafts of fetal rat spinal cord: A Golgi study of neuronal morphology and organization. Exp Neurol 1990; 108:122-9.

56. Jakeman LB, Reier PJ, Bregman BS et al. Differentiation of substantia gelatinosa-like regions in intraspinal transplants of embryonic spinal cord tissue in the rat. Exp Neurol 1989; 103:17-33.

57. Tessler A, Himes BT, Houle J et al. Regeneration of adult dorsal root axons into transplants of embryonic spinal cord. J Comp Neurol 1988; 270:537-48.

58. Henschen A, Hökfelt T, Elde R et al. Expression of eight neuropeptides in intraocular spinal cord grafts: Organotypic and disturbed patterns as evidenced by immunohistochemistry. Neuroscience 1988; 26:193-213.

59. Sobkowicz HM, Guillery RW, Bornstein MB. Neuronal organization in long term cultures of the spinal cord of the fetal mouse. J Comp Neurol 1968; 270:537-48.

60. Naftchi NE, Abrahams SJ, Crain SM et al. Presence of leucine-enkephalin in organotypic explants of fetal mouse spinal cord. Peptides 1981; 2:57-60.

61. Björklund A, Stenevi U, Svengaard NA. Growth of transplanted monoaminergic neurons into the adult hippocampus along the perforant path. Nature 1976; 262:787-90.

62. Björklund A, Stenevi U. Reformation of the severed septohippocampal cholinergic pathway in the adult rat by transplanted septal neurons. Cell Tissue Res 1977; 185:289-302.

63. Björklund A, Stenevi U. In vivo evidence for a hippocampal adrenergic neurotrophic factor specifically released on septal deafferentation. Brain Res 1981; 229:403-28.

64. Björklund A, Segal M, Stenevi U. Functional reinnervation of rat hippocampus by locus coeruleus implants. Brain Res 1979; 170:409-26.

65. Raisman G, Morris RJ, Zhou CF. Specificity in the reinnervation of adult hippocampus by embryonic hippocampal transplants. Prog Brain Res 1987; 71:325-33.

66. Gage FH, Björklund A. Enhanced graft survival in the hippocampus following selective denervation. Neuroscience 1986; 17:89-98.

67. Gage FH, Buzsaki G, Nilsson O et al. Grafts of fetal cholinergic neurons to the deafferented hippocampus. Prog Brain Res 1987; 71:335-47.

68. Bignami A, Dahl D. The astroglial response to stabbing. Immunofluorescence studies with antibodies to astrocyte-specific protein (GFA) in mammalian and submammalian vertebrates. Neuropathol Appl Neurobiol 1979; 2:99-110.

69. Berry M. Regeneration and plasticity in the CNS. In: Swash M, Kennard C, eds. Scientific Basis of Clinical Neurology. Edinburgh: Churchill-Livingstone, 1985:658-79.

70. Reier PJ, Stansaas LJ, Guth L. The astrocytic scar as an impediment to regeneration in the central nervous system. In: Kao CC, Bunge RP, Reier PJ, eds. Spinal Cord Reconstruction. New York: Raven Press, 1983:163-96.

71. Reier PJ, Houle JD. The glial scar: Its bearing on axonal elongation and transplantation approaches to CNS repair. In: Waxman SG, ed. Physiologic Basis for Functional Recovery in Neurological Disease, Adv Neurol. Vol 47. New York: Raven Press, 1988:87-138.

72. Wujek JR, Reier PJ. Astrocytic membrane morphology: Differences between mammalian and amphibian astrocytes after axotomy. J Comp Neurol 1984; 222:607-19.

73. Krüger S, Sievers J, Hansen C et al. Three morphologically distinct types of interface develop between adult host and fetal brain transplants: Implications for scar formation in the adult central nervous system. J Comp Neurol 1986; 249:103-16.

74. Krikorian JG, Guth L, Donati EJ. Origin of the connective tissue scar in the transected rat spinal cord. Exp Neurol 1981; 72:698-707.

75. Houle JD, Reier PJ. Transplantation of fetal spinal cord tissue into the chronically injured adult rat spinal cord. J Comp Neurol 1988; 269:535-47.

76. Reier PJ, Houle JD, Jakeman L et al. Transplantation of fetal spinal cord tissue into acute and chronic hemisection and contusion lesions of the adult rat spinal cord. Prog Brain Res 1988; 78:173-9.

77. Mathewson AJ, Berry M. Observations on the astrocyte response to a cerebral stab wound in adult rats. Brain Res 1985; 327:61-9.

78. Devor M. Neuroplasticity in the rearrangement of olfactory tract fibers after neonatal transection in hamsters. J Comp Neurol 1976; 166:49-72.

79. Kalil K, Reh T. Regrowth of severed axons in the neonatal CNS. Science 1979; 205:1158-61.

80. Bregman BS, Goldberger ME. Anatomical plasticity and sparing of function after spinal cord damage in neonatal cats. Science 1982; 217:553-5.

81. Schneider GE, Jhaveri S, Edwards MA et al. Regeneration, rerouting and redistribution of axons after early lesions. In: Eccles J, Dimitrijevic M, eds. Recent Achievements in Restorative Neurology. Vol 1. Basel: S Karger, 1985:291-310.

82. Silver J, Lorenz SE, Wahlsten D et al. Axonal guidance during development of the great cerebral commissures: Descriptive and experimental studies, in vivo, on the role of preformed glial pathways. J Comp Neurol 1982; 210:10-29.

83. Silver J, Ogawa MY. Postnatally induced formation of the corpus callosum in acallosal mice on glia-coated cellulose bridges. Science 1983; 220:1067-9.

84. Stansaas LJ, Partlow LM, Burgess PR et al. Inhibition of regeneration: The ultrastructure of reactive astrocyted and abortive axon terminals in the transition zone of the dorsal root. Prog Brain Res 1987; 71:457-68.

85. Smith GM, Miller RH, Silver J. Changing role of forebrain astrocytes during development, regenerative failure, and induced regeneration upon transplantation. J Comp Neurol 1986; 251:23-43.

86. Kalderon N, Williams CA. Extracellular proteolysis: Developmentally regulated activity during chick spinal cord histogenesis. Dev Brain Res 1986; 25:1-9.

87. Bernstein JJ, Goldberg WJ. Grafted fetal astrocyte migration can prevent host neuronal atrophy: Comparison of astrocytes from cultures and whole piece donors. Restor Neurol Neurosci 1991; 2:261-70.

88. Nornes H, Björklund A, Stenevi U. Transplantation strategies in spinal cord regeneration. In: Sladek Jr JR, Gash DM, eds. Neural Transplants Development and Function. New York: Plenum Press, 1984:407-21.

89. Jakeman L, Reier PJ. Axonal projections between fetal spinal cord transplants and the adult rat spinal cord: A neuroanatomical tracing study of local interactions. J Comp Neurol 1991; 307:311-34.

90. Bregman BS, Kunkel-Bagden E, McAtee M et al. Extension of the critical period for developmental plasticity of the corticospinal pathway. J Comp Neurol 1989; 282:355-70.

91. Bregman BS. Development of serotonin immunoreactivity in the rat spinal cord and its plasticity after neonatal spinal cord lesions. Dev Brain Res 1987a; 34:245-63.

92. Bregman BS, Bernstein-Goral H. Both regenerating and late-developing pathways contribute to transplant-induced anatomical plasticity after spinal cord lesions at birth. Exp Neurol 1991; 112:49-63.

93. Bernstein DR, Stelzner DJ. Developmental plasticity of the corticospinal tract (CST) following mid-thoracic "over-hemisection" in the neonatal rat. J Comp Neurol 1983; 221:371-85.

94. Bregman BS, Goldberger ME. Infant lesion effect: Anatomical correlates of sparing and recovery of function after spinal cord damage in newborn and adult cats. IIIrd ed. Dev Brain Res 1983c; 9:137-54.
95. Kalil K, Reh T. Light and electron microscopic study of regrowing pyramidal tract fibers. J Comp Neurol 1982; 211:265-75.
96. Bregman BS, Goldberger ME. Infant lesion effect: Effect of neonatal spinal cord lesion on motor development in the kitten. Ist ed. Dev Brain Res 1983a; 9:103-17.
97. Bregman BS, Goldberger ME. Infant lesion effect: Sparing and recovery of function after spinal cord damage in newborn and adult cats. IInd ed. Dev Brain Res 1983b; 9:119-35.
98. Diener PS, Bregman BS. Fetal spinal cord transplants support growth of supraspinal and segmental projections after cervical spinal cord hemisection in the neonatal rat. J Neurosci 1998; 18:779-93.
99. Diener PS, Bregman BS. Fetal spinal cord transplants support the development of target reaching and coordinated postural adjustments after neonatal cervical spinal cord injury. J Neurosci 1998; 18:763-78.
100. Bregman BS. Spinal cord transplants permit the growth of serotonergic axons across the site of neonatal spinal cord transection. Dev Brain Res 1987b; 34:265-79.
101. Bregman BS, Reier PJ. Neural tissue transplants rescue axotomized rubrospinal cells from retrograde death. J Comp Neurol 1986; 244:86-95.
102. Sims TJ, Gilmore SA, Waxman SG et al. Dorsal-ventral differences in the glia limitans of the spinal cord: An ultrastructural study in developing normal and irradiated rats. J Neuropathol Exp Neurol 1985; 44:415-29.
103. Caroni P, Schwab ME. Antibody against myelin-associated inhibitor of neurite growth neutralizes nonpermissive substrate properties of CNS white matter. Neuron 1988; 1:85-96.
104. Schnell L, Schwab ME. Axonal regeneration in the rat spinal cord produced by an antibody against myelin-associated neurite growth inhibitors. Nature 1990; 343:269-72.
105. Carbonetto S, Gruver MM, Turner DC. Nerve fiber growth in culture on fibronectin, collagen and glycosaminoglycan substrates. J Neurosci 1983; 3:2324-35.
106. Carbonetto S, Evans D, Cochard P. Nerve fiber growth in culture on tissue substrata from central and peripheral nervous systems. J Neurosci 1987; 7:610-20.
107. Cornbrooks CJ, Carey DJ, McDonald JA et al. In vivo and in vitro observations on laminin production by Schwann cells. Proc Natl Acad Sci USA 1983; 80:3850-4.
108. Manthorpe M, Engvall E, Ruoshlahti E et al. Laminin promotes neuritic regeneration from cultured peripheral and central neurons. J Cell Biol 1983; 97:1882-90.
109. Sosale A, Robson JA, Stelzner DJ. Laminin distribution during corticospinal tract development and after spinal cord injury. Soc Neurosci Abs 1987; 13:749.
110. Bregman BS, Kunkel-Bagden E. Effect of target and nontarget transplants on neuronal survival and elongation after injury to the developing spinal cord. Prog Brain Res 1988; 78:205-11.
111. Kunkel-Bagden E, Bregman BS. Spinal cord transplants enhance the recovery of locomotor function after spinal cord injury at birth. Exp Brain Res 1990; 81:25-34.
112. Reier PJ, Stokes BT, Thompson FJ et al. Fetal cell grafts into resection and contusion/compression injuries of the rat and cat spinal cord. Exp Neurol 1992; 115:177-88.
113. Goldberger ME. The use of behavioral methods to predict spinal cord plasticity. Restor Neurol Neurosci 1991; 2:339-50.
114. Stokes BT, Reier PJ. Fetal grafts alter chronic behavioral outcome after contusion damage to the adult spinal cord. Exp Neurol 1992; 116:1-12.
115. Anderson DK, Reier PJ, Wirth III ED et al. Delayed grafting of fetal CNS tissue into chronic compression lesions of the adult cat spinal cord. Restor Neurol Neurosci 1991; 2:309-25.
116. Terada N, Hamazaki T, Oka M et al. Bone marrow cells adopt the phenotype of other cells by spontaneous cell fusion. Nature 2002; 416:542-5.
117. Cao QL, Zhang YP, Howard RM et al. Pluripotent stem cells engrafted into the normal or lesioned adult rat spinal cord are restricted to a glial lineage. Exp Neurol 2001; 167:48-58.
118. McDonald JW, Liu XZ, Qu Y et al. Transplanted embryonic stem cells survive, differentiate and promote recovery in injured rat spinal cord. Nat Med 1999; 5:1410-2.
119. Teng YD, Lavik EB, Qu X et al. Functional recovery following traumatic spinal cord injury mediated by a unique polymer scaffold seeded with neural stem cells. Proc Natl Acad Sci USA 2002; 99:3024-9.

CHAPTER 6

Replacement of Specific Neuronal Populations in the Spinal Cord*

Antal Nógrádi

Introduction

As discussed in Chapter 5, embryonic spinal cord grafted into the injured spinal cord of neonatal and adult animals can serve as a relay tissue bridge for axonal growth and regeneration, and promote some degree of functional recovery in the host. In these experiments the pattern of host axonal projections to the grafts has been well characterized, but less emphasis has been placed on the efferent connections of the graft. Furthermore, much fewer attempts have been made to look more selectively at the behaviour of specific subpopulations of grafted neurons. At the time of transplantation embryonic spinal cord contains a mixed population of already committed neurons and neuronal precursors which might respond differently to transplantation. In this chapter we will discuss a different experimental approach which explores the possibility of replacing a defined population of the host neurons by homologous cells present in a graft. This approach, focusing on embryonic neurons of a particular type, also provides an interesting opportunity to follow up their fate after grafting.

Motoneurones are large cholinergic neurons localized to the motor nuclei of the brain stem and spinal cord, and they are the first neuronal population generated during development of the spinal cord.[1] They are also the only central neurons which extend their axons outside the CNS to innervate a nonneuronal target - skeletal muscle. Their selective loss is a hallmark of certain human neurological disorders like spinal muscular atrophies, adult motoneurone disease and poliomyelitis. These disorders, where the loss of motoneurones is the most important cause of the disability underline the significance of studies focusing on grafting motoneurones into a spinal cord which has previously been depleted of its motoneurones. It remains a tantalizing possibility that some results of this research could be relevant not only to our understanding of pathology, but also to treatment of neurological disorders affecting motoneurones.

In the late 1980s different experimental models of motoneurone transplantation were introduced. A number of questions had to be addressed at the beginning of this research. First, what happens to the grafted motoneurones which, as the earliest-born neurons, either are already present in the grafts of embryonic spinal cord, or are due to be born shortly after grafting. It was noted in earlier studies that large motoneurone-like cells were relatively underrepresented in the grafts of embryonic spinal cord, especially if the grafted tissue was obtained from older

---

*Based on chapter in previous edition written by Katarzyna Sieradzan and Gerta Vrbová.

Transplantation of Neural Tissue into the Spinal Cord, Second Edition,
edited by Antal Nógrádi. ©2006 Eurekah.com and Springer Science+Business Media.

embryos. This finding could be explained by their selective death or arrested maturation which rendered them indistinguishable from the other types of grafted neurons.

However, the experiments of Sotelo and Alvarado-Mallart on transplantation of Purkinje cells into the cerebellar cortex of mice[2,3] suggested yet another possibility. In their studies fragments of cerebellar primordia or suspensions of embryonic cerebellar neurons were grafted into the cerebellar cortex of mutant mice with a hereditary ataxia due to the Purkinje cell degeneration (Purkinje Cell Degeneration or PCD mice). In this strain of mice the cerebellar cortex develops normally but about three weeks postnatally the Purkinje cells degenerate completely. Thus, this animal model provides an unique opportunity to study what happens when developing Purkinje cells from the graft are confronted with a mature cortex depleted of the homologous highly specialized neurons. Some Purkinje cells left the grafts and migrated tangentially into the host cortex settling in the upper third of the molecular layer. These cells subsequently developed a typical sagitally oriented dendritic tree and made synaptic contacts with appropriate host afferents. Few of them also extended their axons into their normal efferent projection field in the dentate nucleus. Thus, the graft-derived Purkinje cells were able to replace missing host neurons in a highly organized, point-to-point connected circuitry. These results suggested a possibility that perhaps some embryonic motoneurones could also migrate from the grafts of embryonic spinal cord into the host grey matter to replace the missing motoneurones of the host. If this was true then the next question was whether the grafted motoneurones would be able (a) to integrate into the host neuronal network, and (b) will they be able to innervate a skeletal muscle of the host.

Conditions for Survival of Transplanted Embryonic Motoneurones

Several factors concerning the graft and the host can influence the survival of the grafted embryonic motoneurones. Selection of the appropriate gestational age of the donor embryos is of obvious importance. Developmental studies using ^{3}H thymidine revealed that motoneurone pools of the rat spinal cord are formed between embryonic day 11 and 13 (ED11 and ED13) (the day of insemination was counted as ED-1).[1,4] In the rat the first motor axons extend outside the spinal cord towards their target muscles on ED 12.[5] Therefore the best time to obtain embryonic spinal tissue for transplantation of motoneurones should be after the onset of motoneurone proliferation but before extension of motor axons to the periphery, i.e.,between 11 and 13 days of gestation.

Some characteristic features of developing motoneurones should also be taken into consideration. During the earliest stages of development of the neuromuscular system motoneurones and mesenchymal cells giving rise to muscle primordia develop independently of each other. However, subsequently motoneurones and their target muscles become dependent on each other and together proceed through a sequence of events ultimately leading to formation of adult type motor units.

Similarly to other developing neuronal populations the final number of cells within the motor pools is a net result of the initial mitotic proliferation of neuroblasts in the germinal zone of the neural tube and naturally occurring cell death, recently attributed to apoptosis of "excess" cells.[6] In mice this naturally occuring motoneurone death begins on ED13 and between ED13 and ED18 67% of cells initially present in the lateral motor columns die.[7] In the rat the timing and magnitude of naturally occurring motoneurone death is similar and seems to be completed before birth.[8]

It has been recognized for a long time that the onset of naturally occurring cell death of motoneurones coincides with the arrival of the motor axons to the muscles.[9-12] The magnitude of motoneurone death can be influenced by experimental manipulations leading to either reduction or expansion of the peripheral field of innervation. For example, removal of a limb bud of a chick embryo at a very early stage of development (ED2.5) does not affect proliferation of neuroblasts and migration of motoneurones to the mantle layer. However, later most motoneurones which would normally have innervated the missing limb died.[13] By contrast, expansion of the peripheral field of innervation by implantation of a supernumerary limb bud can prevent death of some, but not all, motoneurones normally destined to die.[14] This dependence on target for survival extends into the early postnatal period. In the neonatal rat, for example, ablation of the limb muscles causes massive death of motoneurones.[15] Similarly, sciatic nerve crush at birth invariably produces death of a large proportion of motoneurones.[16,17] Identical injury to the sciatic nerve inflicted on postnatal day 5 (PD 5) does not induce cell death indicating that, by that stage, motoneurones have acquired the ability to survive temporary isolation from the muscle. It is possible that the disastrous consequences of a peripheral nerve injury during the critical period of development are due to direct axotomy.[18] However, a similar degree of motoneurone death can be induced by preventing neuromuscular transmission with a postsynaptic blocking agent.[19] These experiments, in which direct surgical or pharmacological damage to motor axons was avoided, prove that motoneurone death after a neonatal nerve injury is, indeed, caused by the lack of interaction with the target during the critical period of development.

It can be expected that embryonic motoneurones grafted into the spinal cord could express a similar sensitivity to deprivation of target muscles. It has to be emphasized that so far no systematic studies addressing this issue have been reported. In some experimental models attempts have been made to give the grafted motoneurones a chance to innervate a surrogate target: one of the muscles of the host.[20] Regardless of these purely neurobiological considerations, this approach could also satisfy the restorative purpose of motoneurone transplantation by showing that the grafted cells are capable of forming functional efferent connections with the host muscles.

If the grafted embryonic motoneurones are to innervate the muscles of the host their axons have to be able to exit the spinal cord of the host. Ideally, the best results could be expected if the axons of grafted motoneurones were able to use the existing anatomical routes like the ventral roots and peripheral nerves. However, studies of axonal regeneration from adult motoneurones axotomized close to their perikarya (within the spinal cord) have shown that regeneration into the ventral roots does not occur unless the roots are cut or avulsed and reimplanted directly in the grey matter.[21,22] As discussed in Chapter 5, the environment of the adult CNS, and in particular that of the white matter, permits only abortive regeneration of axons of the central neurons. On the other hand, peripheral nerves implanted into the grey matter of the brain or spinal cord attract axons from the neighbouring neurons and provide an excellent environment for their long-distance (up to 30 mm in the rat) regeneration (reviewed by Aguayo,[23] 1985; Bray[24] et al, 1987). Thus, even if the axons from the grafted neurons would have failed to exit the spinal cord using the existing routes, they still could be led out towards the periphery using a conduit of a nerve implanted in the vicinity of the graft.

The environment of the host spinal cord also influences the survival of grafted motoneurones. Many more grafted cells survive in a spinal cord where the host motoneurones have been depleted. However, since the main interest in grafting embryonic motoneurones into an adult spinal cord is to replace lost cells this factor may be an advantage. Thus the 3 factors that seem to determine the survival of grafted cells are: (i) depletion of host motoneurones, (ii) availability of a muscle that can be reinnervated via a peripheral nerve and (iii) the nature of the peripheral nerve implant. In the following sections the role of these factors is discussed.

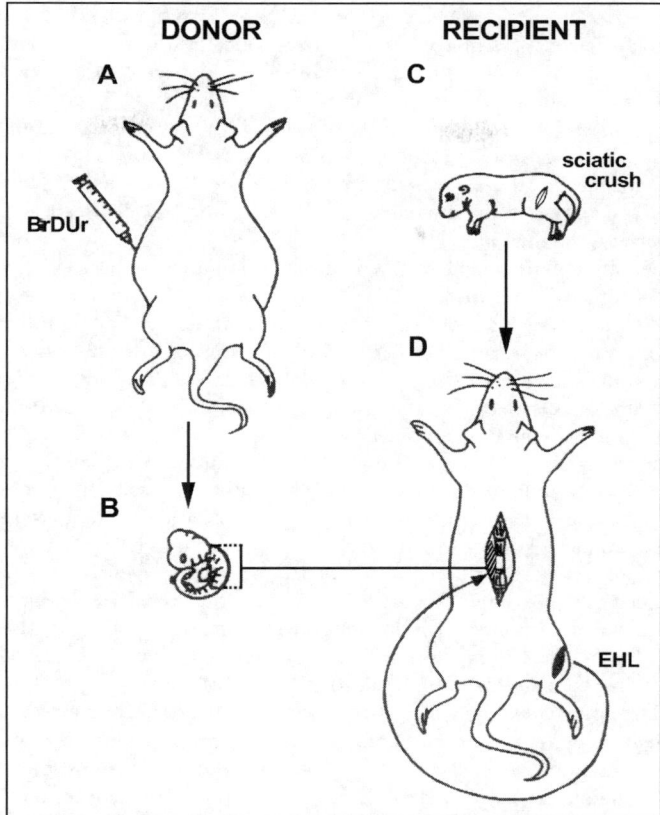

Figure 1. The experimental design is illustrated. The motoneurones of the donor embryos were prelabelled with BrdU administered intraperitoneally to the pregnant female on ED10. One to two days later the embryonic spinal cord was dissected and used for transplantation. The future hosts had their sciatic nerve crushed at birth. The embryonic graft was placed into the motoneurone-depleted L4-L5 lumbar segments. During the same procedure a muscle, with a length of its nerve attached to it, was dissected from the contralateral hindlimb and transposed alongside the vertebral column. The free end of the nerve was implanted in the spinal cord at the site of the graft to form a conduit for axonal elongation. Reproduced with permission from ref. 20. Sieradzan K, Vrbová G. Neuroscience 1989; 31:115-130.

Transplantation of Embryonic Motoneurones into the Adult Spinal Cord

Establishing Optimal Conditions for Survival of Grafted Embryonic Motoneurones in an Adult Host

Attempts of grafting of embryonic motoneurones were carried out in the spinal cord of adult rats. The experimental model used in these studies was designed so as to provide favorable conditions for the survival of the grafted embryonic motoneurones[20] i.e., the host cord was depleted of its motoneurones and a skeletal muscle with a nerve conduit provided (see Fig. 1). A small fragment of the ventral part of the spinal cord from ED11-ED12 embryos was inserted into the lumbar segments of the host spinal cord (L4-L5). The host spinal cord's sciatic motor pool had been selectively depleted several weeks earlier by subjecting the future hosts to sciatic nerve crush at birth. This injury is known to produce death of sciatic motoneurones distributed at L4-L5 level.[16] Another experimental manipulation used in this model was to supply the grafted motoneurones with a target muscle which they could innervate. One of the host's muscles (Extensor Hallucis Longus {EHL} or Soleus {SOL}) was dissected from the contralateral hindlimb of the host with a length of its nerve attached to it. This neuromuscular preparation was attached to the paravertebral muscles. The cut end of the nerve was then inserted into the spinal cord in the immediate proximity of the graft to create a conduit for axonal growth towards the muscle.

The motoneurones of donor embryos were prelabelled in utero with a synthetic analogue of thymidine - 5-bromo-2'-deoxyuridine (BrdU) administered in pulsed injections to a pregnant female on ED10, i.e.,during proliferation of lumbar motoneurone pools in the embryos. The tissue was harvested for grafting one to two days later. Similarly to ^{3}H-thymidine, BrdU labels DNA of dividing cells and can be used to mark birth dates of neurons. The label can be conveniently visualized in the nucleus by immunocytochemistry with the anti-BrdU antibody,[25] allowing identification of embryonic motoneurones in the host spinal cord. Several weeks after transplantation viable and well integrated grafts could be found in the spinal cords of the majority of host animals. In light microscopy the grafts had typical cellular composition and organization, well described in earlier studies (see Chapter 5). Examining the grafts one is often confronted with a problem of deciding which cells fulfil the criteria for a motoneurone. Among numerous smaller neurons the grafts contain some larger perikarya, which on Nissl preparations resemble small motoneurones (they are usually smaller than typical alpha-motoneurones in the host ventral horn). However, a specific motoneurone marker is needed to ascertain the nature of these cells. There are a few possible markers, for example (i) the labelling cells according to their birth date with BrdU or a similar substance, or (ii) the expression of the enzymes involved in cholinergic transmission, like acetylcholinesterase (AChE) or choline acetyltransferase (ChAT). Although these markers are not entirely specific, because a number of other spinal neurons also have cholinergic phenotype,[26,27] the motoneurones are the most numerous subpopulation of the cholinergic cells in the spinal cord. More specific markers for motoneurones have recently been described, but these were not available when the first grafting experiments were carried out. Typically, AChE histochemistry reveals variable numbers of AChE-positive neurons, which sometimes form well defined clusters strikingly resembling a ventral horn of the developing spinal cord[28] (Fig. 2). It has been proposed that at least a proportion of these cells could be motoneurones which survived transplantation but became arrested in their maturation.[28] The ventral horn-like areas can be seen in some grafts but are by no means a regular feature. These appearances may depend on a number of technical variables, such as dissection of embryonic tissue including different proportions of the ventral spinal cord, or on the orientation of the tissue during transplantation.

Interestingly, it was observed that the large AChE positive neurons appeared to be more frequent in occasional "bad" experiments, where the graft was unintentionally contaminated with mesenchymal cells, including muscle (Sieradzan and Vrbová, personal unpublished observations). In these cases mature muscle fibres, fat and connective tissue developed intraspinally in the vicinity, or even within, the graft. These muscle fibres often had well differentiated end-plates which stained for AChE. This finding is consistent with recent findings where motoneurones grafted into a denervated skeletal muscle or peripheral nerve are able to survive and establish connections with the host muscle fibres.[29,30]

One way of distinguishing embryonic motoneurones from both other cell types present in the grafts, and from neurons of the host, is to prelabel them with BrdU. Immunocytochemistry against BrdU revealed a typical pattern of distribution of the labelled nuclei of embryonic cells in the host spinal cord (see Fig. 3). Within the grafts variable numbers of labelled cells were

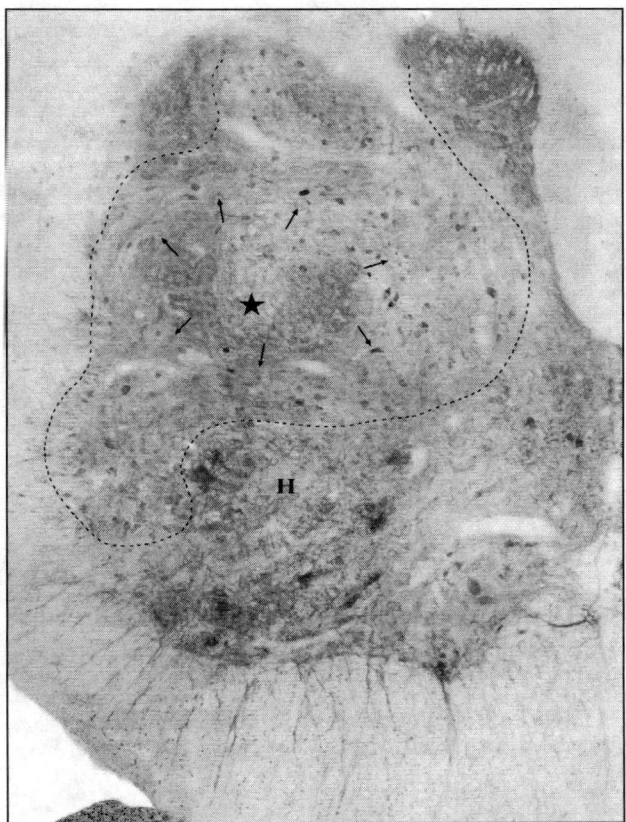

Figure 2. A composite figure of a transverse section through the spinal cord containing a large embryonic graft is shown. The section was reacted histochemically for acetylcholinesterase (AChE). The interface between the graft (*) and the host (H) is marked with broken line. Note the presence of AChE-positive neurons within the graft while the centre of the graft is free of AchE positive neurons. In the ventral horn of the host cord the typical large and intensely stained motoneurones are surrounded by a dense plexus of AChE-containing fibres.

dispersed in a random fashion. The BrdU positive nuclei were of different sizes and varied in intensity of staining for BrdU. In contrast, the cells with nuclei heavily labelled with BrdU were often distributed on the graft-host interface, particularly in the areas of apposition to the host ventral horn. Moreover, some of these intensely labelled nuclei were found outside the grafts in the adjacent grey matter of the host, most often in the ventral horn. The intense labelling of these cells with BrdU showed that their birth-date coincided with administration of the marker, i.e.,they were motoneurones from the embryonic grafts. Indeed, some of these heavily labelled nuclei were found to belong to the cell bodies with motoneurone-like morphological features.

Thus, at least a proportion of embryonic motoneurones present in the grafts at the time of transplantation survived in the host spinal cord and migrated from the graft into the host grey matter showing a preference to the motoneurone-depleted ventral horn.[20,31,33] Thus, after a long interval after depletion of the host motoneurone pool, when the spinal circuitry had

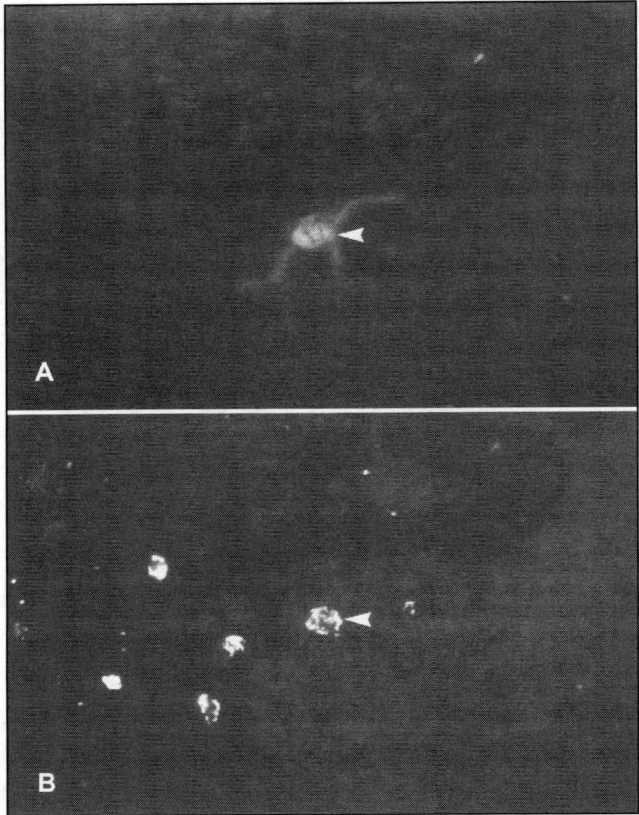

Figure 3. A,B) Example of a neuron double-labelled (arrowhead) with the retrograde tracers FB and DY injected into the reinnervated EHL muscle (A), and with the embryonic marker BrdU (B). Note that the graft contains several BrdU positive nuclei in this section.

probably been extensively remodelled, selective depletion of the host ventral horn promoted survival of the homologous embryonic neurons. These results were obtained from animals that had their a considerable proportion of their own motoneurones depleted at birth. In animals that had their normal number of motoneurones the survival of grafted motoneurones was very poor.[31]

Attempts have also been made to graft cell suspensions enriched in motoneurones. Cultured motoneurones were grafted into the intact spinal cord and the intact brain.[32,33] Surprisingly, only 5% of the grafted motoneurones survived, and some of them attached to the ependyma of cerebral ventricles. These embryonic motoneurones migrated for up to 2 mm from the site of injection both in the host grey and white matter and appeared to be relatively small and immature.

Later studies of motoneurone-enriched cell suspensions,[34] or purified motoneurone fractions cultured and labelled in vitro prior to grafting,[35] showed that some of these cells express AChE and CGRP, but expression of ChAT was generally poor, similarly to the solid grafts of embryonic spinal cord.

Innervation of Host Muscles by the Grafted Motoneurones

As described above, grafted embryonic neurons were able to innervate a host muscle by extending their axons along the peripheral nerve conduit coimplanted in the host spinal cord. Using retrograde labelling from the muscle implant, many labelled cells were found in the host tissue, and majority of these cells were motoneurones. The majority of cells were at the graft-host interface. Retrograde labelling of peripheral nerve implants that had no access to muscle, when inserted into the cervical spinal cord of adult rats showed extensive growth of axons from all types of neurons. However the numbers of axons from grafted motoneurone-like cells were small.[36] Since the origin of the cells was not examined in these experiments some of the retrogradely labelled neurons found in the host grey matter could be motoneurone-like cells of embryonic origin which migrated from the graft.

To provide a more detailed description of the origin of retrogradely labelled cells a double-labelling method was used. Neurons retrogradely labelled with fluorescent tracers from the target muscle were predominantly distributed in the grey matter of the host and about 15% of the total number of these cells contained BrdU in their nuclei confirming their embryonic origin.[37] These findings indicate that a small number of the BrdU-positive embryonic motoneurone-like cells not only migrated from the graft into the host ventral horn, but also extended their axons outside the host spinal cord to innervate the target muscle.

Morphological studies of the target host muscles showed features consistent with denervation and reinnervation. There was a variability of muscle fiber sizes with some proliferation of connective tissue, and fibre type grouping was detected on histochemical staining. Most importantly the muscles were functionally reinnervated; since strong contractions similar to those produced by the original muscle in its usual position could be elicited by electrical stimulation applied to the nerve bridge.[20] Numerous motor axons contacting well developed end-plates were visualized by acetylcholinesterase-silver stain. Thus, despite a profound developmental mismatch between grafted mebryonic motoneurones and the target muscle, the embryonic motoneurones were able to interact with the adult denervated muscle and could be of potential functional value.

In a recent study some of these results obtained on the lumbar cord of adult rats were confirmed in the more rostral cord. Motoneurone enriched grafts from embryos were transplanted into the cervical cord of adult rats and connected to the musculocutaneous nerve. A large proportion of these motoneurones extended their axons to the biceps muscle and like in the lumbar cord established connections with the denervated biceps muscle.[38]

Reinnervation of an Avulsed and Reimplanted Lumbar Ventral Root by Axons of Grafted Motoneurones

Although axons of grafted neurons were able to enter the nerve conduit to a paravertebrally placed hindlimb muscle (EHL or SOL) they were unable to enter the lumbar ventral roots.[39-41] However, it has been shown that axons of injured adult motoneurones can enter a ventral root that had been avulsed at the exit from the spinal cord and then reimplanted into the ventral horn.[21,22,42] Such an avulsed ventral root reimplanted close to the graft is also an excellent conduit for axons of grafted motoneurones for many motoneurones of embryonic origin extended their axons into the reimplanted ventral root.[43] Interestingly, many more motoneurones entered the reimplanted ventral root than the EHL nerve-muscle implant. Approximately 20% of these reinnervating neurons was confirmed to be of embryonic origin with BrdU labelling[43] and 83% of them proved to be motoneurone (Nógrádi and Vrbová, unpublished results). At least 75% of the neurons innervating the reimplanted ventral root reached and reinnervated the hindlimb muscles. The reinnervation of the denervated hindlimb muscles was proven both histologically and functionally (Fig. 4). Animals that had their L4 ventral root

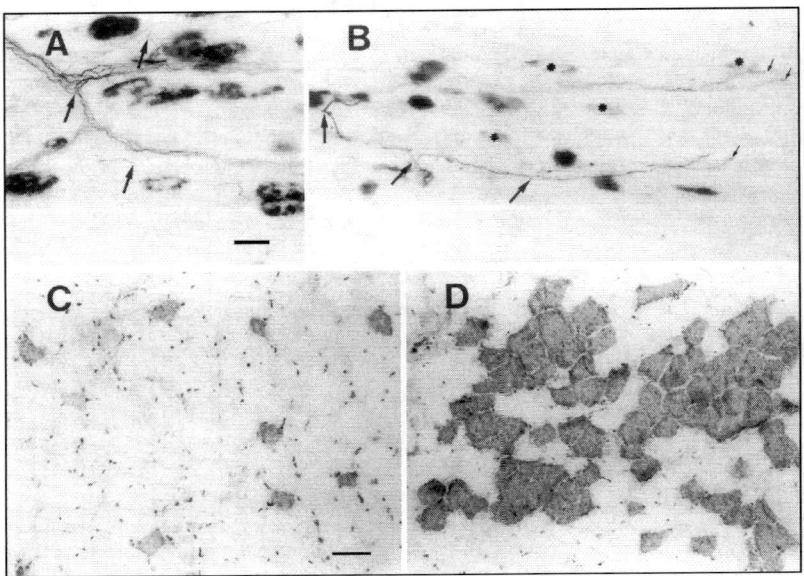

Figure 4. A,B) A longitudinal section through the EDL muscle reinnervated by axons of host and grafted neurons via the avulsed and reimplanted L4 ventral root. The section has been processed using acetylcholinesterase-silver stain. A) Note the numerous axons and end plates. Arrows point to sprouting axons. B) Denervated end plates (*) are intermingled with some reinnervated end plates. Large arrows point to the numerous sprouting axons, small arrows indicate the pathfinding growth cones at the end of axons 5 month after grafting. C,D) Cross sections of intact (C) and reinnervated (D) EDL muscles. The muscles were reinnervated via the reimplanted L4 ventral root and processed for slow myosin immunohistochemistry. D) Note the grouping of slow myosin positive muscle fibres in the reinnervated EDL muscles. Scale bars = 25 μm in A-D.

avulsed and reimplanted without an embryonic graft were severely handicapped, but those that had embryonic grafts walked without any major visible deficit.

Apart from the findings on the reinnervation of hindlimb muscles, immunocytochemical analysis of the grafted cells showed that considerably higher number of grafted neurons expressed ChAT[44] and CGRP than in experiments where a small nerve-muscle implant was connected to the graft. This made it possible to perform double labelling on the retrogradely labelled neurons with one of these markers. Almost all the CGRP positive cells in the graft were retrogradely labelled, although a considerable number of reinnervating cells did not contain the marker, suggesting that not all the reinnervating neurons reached the same level of maturity and differentiation.

Transplantation in Models of Neurodegeneration

Indiscriminate degeneration of neurons can be induced by injection of high doses of kainic acid. Grafts of cell suspensions from embryonic spinal cord into such neurone-depleted regions formed a well defined "neonucleus" repopulating the lesioned area.[45,46] The transplants contained mixed neuronal populations with small, medium and few large-size somata which resembled motoneurones. Presumed motoneurones did contain CGRP in their cytoplasm, although expression of the peptide was less intense than in the host motoneurones in the adjacent intact ventral horn. Similarly to the findings in other experimental models, this feature could reflect abnormal maturation and/or connectivity of the grafted motoneurones.

A selective depletion of adult motoneurones with neurotoxic lectins provides a relatively good experimental model of human motoneurone disease of adult onset.[47,48] Following injection of a neurotoxic lectin, volkensin or ricin into the muscle or peripheral nerve, the lectin is retrogradely transported to motoneurone somata causing their death, but there is no evidence of transsynaptic spread of the toxin.[49] Embryonic grafts placed into these motoneurone-depleted areas behaved in a similar manner to that seen in the spinal cord depleted of motoneurones by neonatal nerve injury. The grafted motoneurone-like cells labelled with BrdU were predominantly localized to the graft-host interface. Some of them also migrated into the host grey matter and were typically found near the junction with the host ventral horn. In this experimental situation the BrdU labelled neurons remained confined to the graft and little migration into the host tissue occurred.[41]

Transplantation of Embryonic Motoneurones into the Spinal Cord of Immature Animals

As discussed in Chapter 5, integration of the embryonic grafts into the spinal cord of young animals appears to be more complete than in the adult. This finding is mirrored by the functional improvement in these young animals. It could therefore be expected that grafted embryonic motoneurones would also survive better in the immature spinal cord. Surprisingly, the opposite seems to be the case.[39,50]

In 5-12 days-old rat pups local depletion of the host motoneurones was induced by neonatal sciatic nerve crush so that at the time of grafting 70-80% died.[16,17] The axons of the grafted motoneurones were given a chance to innervate an autologous muscle implant through a bridge of the peripheral nerve coimplanted in the vicinity of the graft.

Survival and general development of the grafts was satisfactory, but retrograde labelling with fluorescent tracers to the nerve-muscle implant revealed very few reinnervating neurons in the host spinal cords. The muscle implants were severely atrophic and did not show any signs of reinnervation. The negative outcome of these experiments, much worse than in the adult animals, was surprising. There was an almost complete absence of embryonic motoneurone-like cells in the host neuropil and it appeared that neither host motoneurones nor embryonic grafted cells were able to extend their axons into this immature nerve-muscle implant. Whether this failure of grafted motoneurones to survive and extend their axons into the immature graft can be attributed to the unfavourable environment of the less mature spinal cord, or to the nerve muscle conduit, was established next.

Therefore the immature autologous target muscle and nerve conduit was replaced by a similar nerve-muscle implant obtained from adult immunocompatible rats.[50] These adult implants were much more efficient in attracting axonal outgrowth from the spinal neurons. However, double-labelling for the presence of BrdU showed that only about 3% of the neurons that sent their axons to the target muscle were of embryonic origin. Thus, even when the target muscle was accessible, the numbers of surviving motoneurone-like cells derived from the graft were much lower in the developing than in the adult animals.

All these data strongly suggested that, apart from the nonpermissive nature of an immature peripheral nerve bridge, the environment of the developing spinal cord was also responsible for poor survival of the grafted embryonic motoneurones. In the rat spinal cord most of axonal growth and synaptogenesis occur during the first postnatal weeks but the latter is not complete until about 30 days after birth.[51] In such dynamically developing environment the chances for migrating embryonic motoneurones to be exposed to the growing host afferents, many of which utilize glutamate as a neurotransmitter, could be higher than in a relatively quiescent adult spinal cord, and at a critical stage of development excessive exposure to glutamate is toxic to spinal neurons.[52-54] (for review see Lowrie and Vrbová,[17] 1992). Thus, in the developing spinal cord excessive activation by the growing glutaminergic inputs from the host could aggravate the stress already imposed upon the grafted embryonic motoneurones by deprivation of target in critical stage of development.

Most Important Points

The available data shows that a number of grafted motoneurones survive transplantation for prolonged periods of time. The grafted motoneurones have a better chance of survival when the host cord is depleted of it's own motoneurones. The motoneurones of graft origin are able to survive and establish functional connections with adult denervated muscle. It is interesting that reducing this developmental mismatch by grafting embryonic motoneurones into a developing spinal cord reduces the chances of survival of grafted cells. In the adult spinal cord the grafted motoneurone-like cells display preferential migration into the ventral horn of the host.

From the practical point of view it is important to define the conditions which could improve survival of grafted embryonic motoneurones. The importance of making contact with the target muscle for the survival of developing motoneurones has already been discussed. Even if the experimental design creates a possibility of innervating a surrogate target, such as a muscle of the host, this obviously would not have a chance to occur during the most vulnerable stage after grafting. Thus, the presence of the muscle is more likely to be of benefit for maturation of the grafted embryonic motoneurones, and possibly for their long-term maintenance, but not for their immediate survival.

However, the major challenge is how to guide the axons of the grafted motoneurones towards the appropriate muscles of the host. At present the ventral root avulsion and reimplantation procedure makes it possible for the axons of grafted motoneurones to reach the denervated hindlimb muscles and induce functional reinnervation.

References

1. Altman J, Bayer SA. The development of the rat spinal cord. Adv Anat Embryol Cell Biol 1984; 85:1-166.
2. Sotelo C, Alvarado-Mallart RM. Reconstruction of the defective cerebellar circuitry in the adult Purkinje cell degeneration mutant mice by Purkinje cell replacement through transplantation of solid embryonic implants. Neuroscience 1987; 20:1-22.
3. Sotelo C, Alvarado-Mallart RM, Gardette R et al. The fate of grafted embryonic Purkinje cells in the cerebellum of the adult "Purkinje cell degeneration" mutant mouse. I. Development of reciprocal graft-host interactions. J Comp Neurol 1990; 295:165-187.
4. Nornes HO, Das GD. Temporal pattern of neurogenesis in spinal cord of rat. I. An autoradiographic study - time and sites of origin and migration and settling patterns of neuroblasts. Brain Res 1974; 73:121-138.
5. Reynolds ML, Fitzgerald M, Benowitz LJ. GAP-43 expression in developing cutaneous and muscle nerves in the rat hindlimb. Neuroscience 1991; 41:201-211.
6. Oppenheim RW. Cell death during development of the central nervous system. Annu Rev Neurosci 1991; 14:453-501.
7. Lance-Jones C. Motoneuron cell death in the developing lumbar spinal cord of the mouse. Dev Brain Res 4:473-479.
8. Oppenheim RW. The absence of significant postnatal motoneuron death in the brachial and lumbar spinal cord of the rat. J Comp Neurol 1986; 246:281-286.
9. Hamburger V. Cell death in the development of the lateral motor column of the chick embryo. J Comp Neurol 1975; 160:535-546.

10. Hamburger V. The Developmental History of the Motor Unit: The F.O. Schmidt Lecture in Neuroscience. In: Neuroscience Research Programme Bulletin. Cambridge, MA: MIT Press, 1976:1-37.

11. Hamburger V, Oppenheim RW. Naturally occurring cell death in vertebrates. Neurosci Comment 1982; 1(2):39-55.

12. Oppenheim RW, Chu-Wang I-W. Aspects of naturally-occurring motoneuron death in the chick spinal cord during embryonic development. In: Burnstock G, O'Brien RAD, Vrbová G, eds. Somatic and Autonomic Nerve Interactions. Amsterdam: Elsevier Science Publishers BV, 1983:56-107.

13. Hamburger V. Regression versus peripheral control of differentiation in motor hypoplasia. Am J Anat 1958; 102:365-410.

14. Hollyday M, Hamburger V. Reduction of the naturally occurring motor neuron loss by enlargement of the periphery. J Comp Neurol 1976; 170:311-320.

15. Romanes GJ. Motor localization and the effects of nerve injury on the ventral horn cells of the spinal cord. J Anat 1946; 80:117-131.

16. Lowrie MB, Subramaniam K, Vrbová G. Permanent changes in muscles and motoneurones induced by nerve injury during a critical period of development of the rat. Dev Brain Res 1987; 31:91-101.

17. Lowrie MB, Vrbová G. Dependence of postnatal motoneurones on their targets: Review and hypothesis. Trends Neurosci 1992; 15:80-84.

18. Lieberman AR. Some factors affecting retrograde neuronal responses to axonal lesions. In: Bellairs R, Gray EG, eds. Essays on the Nervous System and Festschrift for Professor JZ Young. Oxford: Oxford University Press, 1974:77-105.

19. Greensmith L, Vrbová G. Alterations of nerve-muscle interactions during postnatal development influence motoneurone survival in rat. Dev Brain Res 1992; 69:125-131.

20. Sieradzan K, Vrbová G. Replacement of missing motoneurones by embryonic grafts in the rat spinal cord. Neuroscience 1989; 31:115-130.

21. Cullheim S, Carlstedt T, Linda H et al. Motoneurones reinnervate skeletal muscle after ventral root implantation into the spinal cord of the cat. Neuroscience 1989; 29:725-733.

22. Carlstedt T, Risling M, Linda H et al. Regeneration after spinal nerve root injury. Restor Neurol Neurosci 1990; 289-295.

23. Aguayo AJ. Axonal regeneration from injured neurons in the adult mammalian central nervous system. In: Cotman CW, ed. Synaptic Plasticity. New York: The Guilford Press, 1985:457-484.

24. Bray GM, Villegas-Peres MP, Vidal-Sanz M et al. The use of peripheral nerve grafts to enhance neuronal survival, promote growth and permit terminal reconnections in the central nervous system of adult rats. J Exp Biol 1987; 132:5-13.

25. Miller MW, Nowakowski RS. Use of bromodeoxyuridine-immunohistochemistry to examine the proliferation, migration and time of origin of cells in the central nervous system. Brain Res 1988; 457:44-52.

26. Barber RP, Phelps PE, Houser CR et al. The morphology and distribution of neurons containing choline acetyltransferase in the adult rat spinal cord: An immunocytochemical study. J Comp Neurol 1984; 229:329-346.

27. Phelps PE, Barber RP, Brennan LA et al. Embryonic development of four different subsets of cholinergic neurons in rat cervical spinal cord. J Comp Neurol 1990; 291:9-26.

28. Clowry GJ, Sieradzan K, Vrbová G. Expression of cholinergic phenotype by embryonic ventral horn neurones transplanted into the spinal cord in the rat. Restor Neurol Neurosci 1994; 6:209-219.

29. Thomas CK, Erb DE, Grumbles RM et al. Embryonic cord transplants in peripheral nerve restore skeletal muscle function. J Neurophysiol 2000; 84:591-595.

30. Guangliang J, Yudong G. The observation of transplanted embryonic motoneurones in the denervated muscles of adult rats. Chinese Med J 1998; 111:63-66.

31. Sieradzan K, Vrbová G. Factors influencing survival of transplanted embryonic motoneurones in the spinal cord of adult rats. Exp Neurol 1991; 114:286-299.

32. Demierre B, Martinou J-C, Kato A. Embryonic motoneurones grafted into the adult CNS differentiate and migrate. Brain Res 1990; 510:355-359.

33. Demierre B, Ruiz-Flandes P, Martinou J-C et al. Grafting of embryonic motoneurones into adult spinal cord and brain. Prog Brain Res 1990; 82:233-237.

34. Sieradzan K, Clowry G, Haynes L et al. The ability of isolated embryonic motoneurones grafted into rat spinal cord to express cholinergic markers and innervate skeletal muscle. Restor Neurol Neurosci 1992; 4:220.

35. Peschanski M, Nothias F, Cadusseau J. Is there a therapeutic potential for intraspinal transplantation of fetal spinal neurons in motoneuronal diseases? Restor Neurol Neurosci 1992; 4:227.

36. Horvat JC, Pecot-Dechavassine M, Mira JC et al. Formation of functional endplates by spinal axons regenerating through a peripheral nerve graft: A study in the adult rat. Brain Res Bull 1989; 22:103-14.

37. Clowry GJ, Vrbová G. Observations on the development of transplanted embryonic ventral horn neurones grafted into adult spinal cord and connected to skeletal muscle implants via peripheral nerve. Exp Brain Res 1992; 91:249-258.

38. Duchossoy Y, Kassar-Duchossoy L, Orsal D et al. Reinnervation of the biceps brachii muscle following cotransplantation of fetal spinal cord and analogous periphral nerve into the injured spinal cordof the adult rat. Exp Neurol 2001; 167:329-340.

39. Sieradzan K, Vrbová G. The ability of developing spinal neurons to reinnervate a muscle through a peripheral nerve conduit is enhanced by cografted embryonic spinal cord. Exp Neurol 1993a; 122:232-243.

40. Nógrádi A, Vrbová G. The use of embryonic spinal cord grafts to replace identified motoneurone pools depleted by a neurotoxic lectin, volkensin. Restor Neurol Neurosci 1992; 4:219.

41. Nógrádi A, Vrbová G. The use of embryonic spinal cord grafts to replace identified motoneurone pools depleted by a neurotoxic lectin, volkensin. Exp Neurol 1994; 129:130-141.

42. Carlstedt T, Grane P, Hallin RG et al. Return of function after spinal cord implantation of avulsed spinal nerve roots. Lancet 1995; 346:1323-1325.

43. Nógrádi A, Vrbová G. Improved motor function of denervated rat hindlimb muscles induced by embryoni spinal cord grafts. Eur J Neurosci 1996; 8:2198-2203.

44. Nógrádi A, Vrbová G. The effect of riluzole treatment in rats on the survival of injured adult and grafted embryonic motoneurones. Eur J Neurosci 2001; 13:113-118.

45. Nothias F, Peschanski M. Homotypic fetal transplants into an experimental model of spinal cord neurodegeneration. J Comp Neurol 1990; 301:520-534.

46. Nothias F, Cadusseau J, Dusart I et al. Fetal neural transplants into the area of neurodegeneration in the spinal cord of adult rat. Restor Neurol Neurosci 1991; 2:283-238.

47. de la Cruz RR, Baker R, Delgado-Garcia JM. Behaviour of cat abducens motoneurones following the injection of toxic ricin into the lateral rectus muscle. Brain Res 1991; 544:260-268.

48. Nógrádi A, Vrbová G. The use of a neurotoxic lectin, volkensin, to induce loss of identified motoneurone pools. Neuroscience 1992; 50:975-986.

49. de la Cruz RR, Pastor AM, Delgado-Garcia JM. Long-term effects of selective target removal on brainstem premotor neurons in the adult cat. Eur J Neurosci 1993; 5:232-239.

50. Sieradzan K, Vrbová G. Observations on the survival of grafted embryonic motoneurones in the spinal cord of developing rats. Exp Neurol 1993b; 122:223-231.

51. Weber ED, Stelzner DD. Synaptogenesis in the intermediate grey region of the lumbar spinal cord in the postnatal rat. Brain Res. 1980; 185:17-37.

52. Brenneman DE, Forsythe ID, Nicol T et al. N-Methyl-D-Aspartate receptors influence neuronal survival in developing spinal cord cultures. Dev Brain Res 1990; 51:63-68.

53. Meldrum B, Garthwaite J. Excitatory amino acids and neurodegenerative disease. Trends Pharmacol Sci 1990; 11:379-387.

54. Regan RF, Choi DW. Glutamate neurotoxicity in spinal cord cell culture. Neuroscience 1991; 43:585-591.

CHAPTER 7

Replacement of Specific Populations of Cells:
Glial Cell Transplantation into the Spinal Cord

Antal Nógrádi

Introduction

In recent years an increasing number of results of successful spinal cord transplantation has been reported. Apart from theoretical interest the main aim of these experiments was to find a possible way to improve the consequences of spinal cord injury or neurological disorders affecting the spinal cord. With few exceptions these studies have focused on the survival of neurons, their ability to express specific neurotransmitters and functional improvements achieved by grafting. Glial cells have also been widely used for transplantation into the brain and spinal cord of normal animals and various mutants. A great number of these studies was concerned with the environment provided by glial cells and the putative trophic factors expressed by the grafts. Apart from grafting the two indigenous macroglial cell types of the CNS, the astrocytes and oligodendrocytes, successful attempts have been made to transplant Schwann cells. These experiments showed that glial cells played an important role in the development, regenerative capacity and function of the spinal cord and the possible use of glial cell grafts in demyelinating and degenerative diseases had been suggested. Apart from some mainly theoretical studies concerning the development of glial cells in an alien environment such as the spinal cord, most of the recent investigations revealed several important features of grafted glia, (oligodendrocytes, their precursors and Schwann cells) namely their excessive capability to migrate and to remyelinate dys- or hypomyelinated CNS areas. Astrocytes were shown to migrate long distances along nerve fibres, blood vessels and through the parenchyma, often as far as the full length of the cord. Similar migratory pattern of oligodendrocytes has been reported by Gout et al[1] (1988) in a mouse mutant's (shiverer) spinal cord. This migration of astrocytes is of particular importance since grafted glial cells deposited at a central part of the spinal cord may exert their effect throughout the whole cord soon after grafting. Similarly, grafted oligodendrocytes may remyelinate substantially larger areas by spreading throughout a demyelinating cord. Schwann cells in addition to their extensive capability to migrate encourage lesioned and regenerating axons to extend processes and maintain their myelinating activity within the CNS. They are therefore used for grafting when regeneration of lesioned axons is required.

Apart from macroglia, microglial cells, a subpopulation of parenchymal macrophages have recently been introduced into the potential therapeutical arsenal of glial cells. Although their action seems to be entirely different from that of macroglial cell types, their use suggested some success in helping the recovery of the injured cord.

Transplantation of Neural Tissue into the Spinal Cord, Second Edition,
edited by Antal Nógrádi. ©2006 Eurekah.com and Springer Science+Business Media.

These investigations were aimed to establish a possible use of glial grafts in human diseases, such as in degenerative disorders so as to ameliorate the function of the degenerated areas, improve the regenerative capability of the injured spinal cord and to achieve some remyelination of demyelinating axons. Below the results obtained from grafting various glial cell types into the spinal cord are summarized and their possible therapeutic use is discussed.

Transplantation of Schwann Cells

Spontaneous remyelination after a demyelinating lesion in some diseases involves the presence of myelinating Schwann cells arising from the periphery. Such remyelination by Schwann cells was observed in the periphery of plaques that developed in patients suffering from multiple sclerosis, in brains affected by experimental allergic encephalomyelitis, Theiler's murine encephalomyelitis virus infection or in focal experimental lesions caused by chemical agents, such as lysolecithine or saporin. In these experimental demyelinating lesions Schwann cells first entered the spinal cord via the dorsal root entry zone and the lateral funiculi and readily remyelinated the spinal cord, resulting in functional recovery. Several weeks after remyelination appeared complete by Schwann cells, endogenous oligodendrocytes expelled the Schwann cells and remyelinated the axons previously being myelinated by Schwann cells without lapse in motor function.[2] The progressive replacement of Schwann cells by oligodendrocytes was accompanied by invasion of astrocyte into areas myelinated by Schwann cells (Fig. 1).

This observation of spontaneous Schwann cell remyelination raised the question whether Schwann cells could provide an alternative source of cells to repair lesions in demyelinated areas in the spinal cord.[3] Such a possibility appeared feasible because remyelination of peripheral nerves by Schwann cells occurs more rapidly than remyelination in the CNS by host oligodendrocytes. This is probably due to the fact that adult oligodendrocytes have a limited migratory and myelinating capacity while mature Schwann cells are able to proliferate and remyelinate. However, this aggressive capacity to remyelinate axons in the CNS can be sometimes disadvantageous to certain extent (for details see later). Another encouraging aspect of grafting Schwann cells into lesioned spinal cord was their capability to provide a favourable environment for regenerating axons possibly because they sequester putative trophic factors as well as extracellular matrix growth promoting molecules.[4] Several authors have reported that neurons from the spinal cord or other part of the CNS are able to extend axons for long distances into peripheral nerve grafts and it was thought that regeneration and remyelination within the CNS may be enhanced by providing the adequate PNS-like environment for damaged axons (for details see Chapter 4).

Figure 1. Transverse L5-6 lumbar spinal cord sections at days 75 (A, B, E) and 150 (C, D, F), immunolabeled for astrocytes (GFAP, *green*), Schwann cells (Schwann/2E, *red*), and oligodendrocytes (MAB 1580, *blue*). A, B, Day 75: the previously demyelinated area is now entirely occupied by Schwann cell immunolabeling (S). Schwann cells are now in direct contact with reactive astrocytes (A). C, D, Day 150: oligodendrocytes (Ol) progressed centrifugally and now occupy the medial part of the area formerly populated by Schwann cells. The astrocytes have also advanced to the periphery where they coexist with residual Schwann cells. *dr*, Dorsal rootlet. E, Day 75: EM section taken from the interface between the oligodendrocyte and Schwann cell myelin (similar to *boxed region in* B) shows the juxtaposition of newly oligodendrocyte myelinated axons (01) and Schwann cell myelinated axons (*Sma*). The majority of oligodendrocyte myelination is characterized by thin myelin sheaths (new myelin) and lack of associated cytoplasm or nucleus. One axon myelinated by residual mature oligodendrocyte myelin (*mOl*) is present. F, Day 150: EM section taken from the interface between the oligodendrocyte and Schwann cell myelin (similar to *boxed region in* D). Numerous thinly oligodendrocyte-myelinated axons (*arrows*) of the oligodendrocyte remyelinating region are adjacent to Schwann cell myelinated axons (*Sma*). The *pale circular structure* at the *top right* of the image is a blood vessel. *Sn*, Schwann cell nucleus. Scale bars: (shown in A) A-D, 100 μm; (shown in E) E) 2.4 μm; *f*, 3.7 μm. Reproduced from J. Neuroscience (2000) 20:9215-9223, with kind permission from the Society for Neuroscience (Copyright 2000).

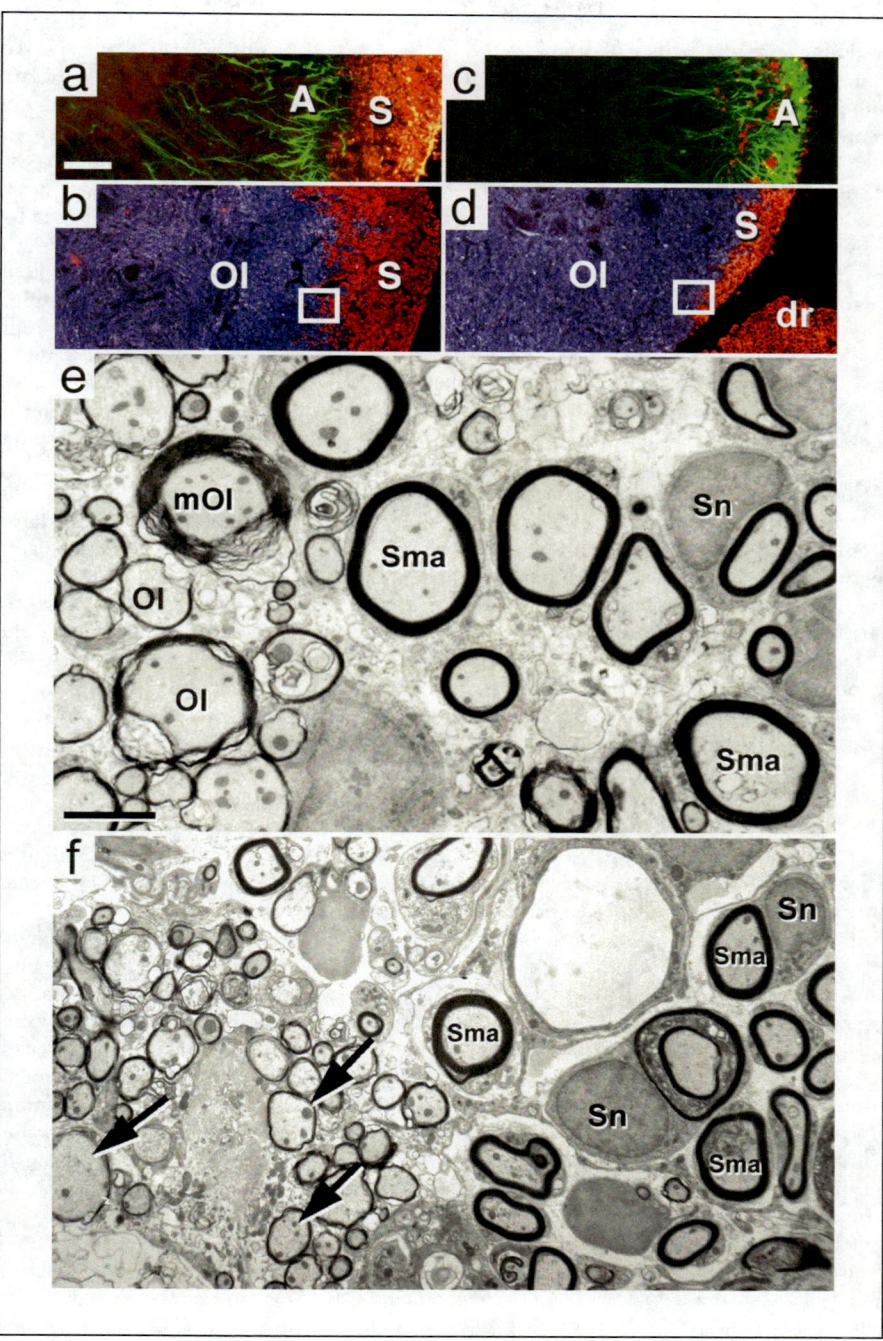

Figure 1. See figure legend on previous page.

Studies by Blakemore and colleagues have shown that experimentally demyelinated spinal cord fibre tracts can be partially remyelinated by Schwann cells arising from a peripheral nerve juxtaposed or inserted into the dorsal columns.[5-8] The autologous nerves need not be placed into the cord, and placing a piece of peripheral nerve into the subarachnoid space[8] or onto the demyelinated dorsal columns also has the desired effect of substantial remyelination.[6] However, the Schwann cell-induced repair never exceeded more than 50% remyelination of naked axons.[8] As the self-repairing process of the spinal cord was suppressed by a high dose of x-ray irradiation it was assumed that all remyelination was due to the graft. Axons remyelinated by Schwann cells could easily be recognised by the signet-ring-like appearance of Schwann cell myelin and the thicker and more compact myelin sheaths. Schwann cells appeared capable to migrate over short distances in response to the appropriate signal, presumably coming from denuded axons. However, this migration was limited and slow and the entry of Schwann cells into the spinal cord was confined to the perivascular spaces which suggested that there might be a significant difference between the extracellular matrices of the CNS and PNS.[8] It has also been shown that although NDFβ (Neu Differentiation Factor β) is an effective mitogen for monkey Schwann cells, but its use does not prevent Schwann cells differentiating into myelinating cells.[9]

Results similar to those obtained by peripheral nerve grafts were obtained by grafts of cultured autologous Schwann cells placed either into chemically demyelinated cords (ethidium bromide,[7,10,11] lysolecithin[9,12] or diphtheria toxin[13]) or into myelin-deficient mutants (quaking mouse[12]). Focal Schwann cell injections into areas of demyelination resulted in limited migration and myelination pattern[12-14] and again, the remyelination was closely related to vessels or demyelinated areas which contained astrocytes. The extent of remyelination is related to the number of grafted Schwann cells and the proliferation of Schwann cells through the lesion rather than to their extensive migration. Extensive remyelination was reported only if

    A. transplantation was performed as early as two days after inducing the demyelination[14] when the astrocytic complement of the lesioned spinal cord was still present,

    B. Schwann cells were grafted directly into the lesion[15] and

    C. grafted Schwann cells were only minimally contaminated with fibroblasts.[16]

Normally astrocytes prevent the intrusion of Schwann cells into the spinal cord, but astrocytic populations altered by x-ray irradiation are believed to provide a surface for migratory Schwann cells and produce extracellular matrix components. In areas where no astrocytes were present only the perivascular collagen was able to promote the migration of Schwann cells and in the absence of structural extracellular matrix components Schwann cells could not migrate alongside naked axons, but formed clumps without forming myelin.[14] On the other hand, unlike astrocytes of an irradiated spinal cord,[7] intact astrocytes limit peripheral remyelination by grafted Schwann cells.[12,13,17] The interaction between astrocytes and unmyelinating Schwann cells does not seem to be so important as that between astrocytes and grafted oligodendrocytes (see: Transplantation of oligodendrocytes) and aggressively myelinating Schwann cells often displaced astrocytes from the areas of demyelination.[7,18] Another aspect of the remyelination process was the competitive relationship between oligodendrocytes and Schwann cells. Schwann cells remyelination is a much quicker process than that produced by oligodendrocytes[7,18] though they do not seem to be mutually inhibitory since they can be found together in areas of remyelination.

Following rapid remyelination such as in demyelinated cat spinal cords remyelinated by Schwann cells Blight and Young[18] observed a faster recovery of some electrophysiological functions (such as cortical somatosensory evoked potentials, CSEP) than in cords repaired by oligodendrocytes. However, in the long-term, when remyelination was complete there was no electrophysiological difference between oligodendrocyte and Schwann cell-remyelinated cords and the extent of recovery was strictly related to the axon survival. Based on these morphological

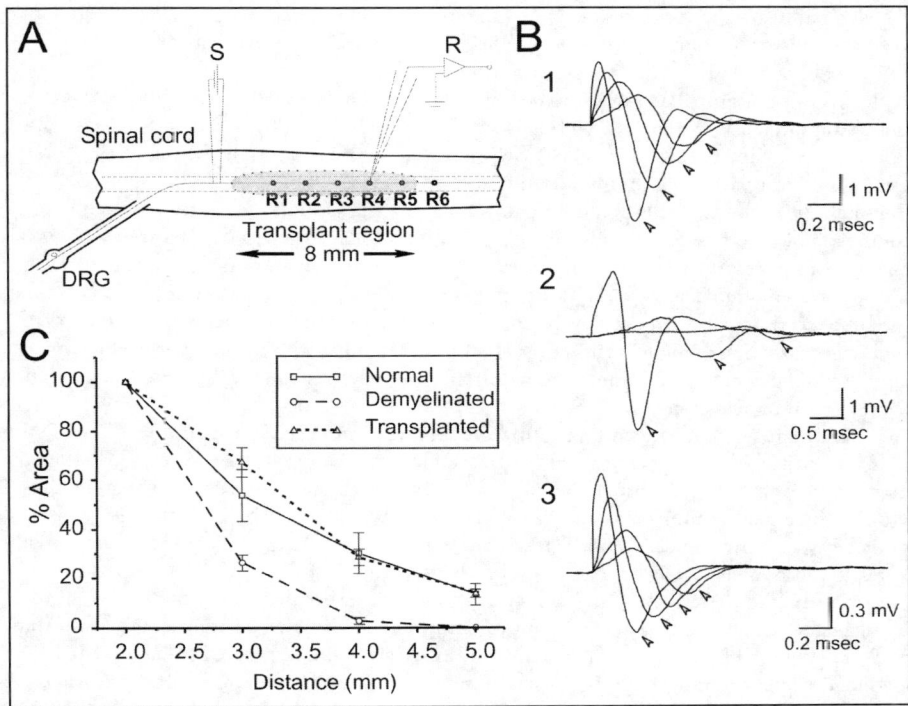

Figure 2. A) Schematic showing the dorsal surface of spinal cord with the positions of the stimulating (S) and recording (R) electrodes. *Shaded region* indicates the area of demylination or remylination. B) Compound action potentials recorded at 1 mm increments along the dorsal columns in control (1), EB-X-demyelinated (2), and transplant-induced remyelinated (3) axons. C) Compound action potential area (*% Area*) plotted versus conduction distance for normal, demyelinated and transplant-induced remyelinated dorsal columns. (n=5). Reproduced from J Neuroscience 1996; 16:3199-3208, with kind permission from the Society for Neuroscience (Copyright 2000).

and physiological findings it became clear that Schwann cells are able of functional remyelination but this does not necessarily mean complete restoration of axonal function.[18] When Schwann cells were co-grafted with astrocytes, remyelination following grafting was near complete and the morphological signs were accompanied by the following electrophysiological features:

    A. restoration of conduction through the lesion,

    B. the remyelinated axons showed enhanced impulse recovery to paired-pulse stimulation and

    C. could fire at higher frequencies compared with both demyelinated and control axons (Fig. 2).[19]

Schwann cells in culture are able to maintain their remyelinating capacity so that they are able to myelinate xenogeneic hosts as well.[11,12,17,20] However, with time Schwann cells appeared to regain their histocompatibility antigens and when grafted into an in vivo environment the remyelinating cells were rejected by 30 days after transplantation[20] or after discontinuing immunosuppression therapy,[12,17] leaving behind demyelinated areas. These results indicated that remyelination was solely due to transplanted cells. Electropysiological analysis has shown that human Schwann cells grafted into a demyelinated rat spinal cord induced functional remyelination suggesting that the conduction block has been overcome.[19]

The pathways and distances taken by migrating Schwann cells were also characterized by various authors.[21-24] Immortalized and purified Schwann cells transplanted at distances from the demyelinating lesion of shiverer and normal spinal cords showed extensive migration towards the lesion. Baron-Van Evercooren et al[23] labelled Schwann cells with a fluorescent dye whilst Langford and Owens[24] used a β-galactosidase gene labelling by infecting Schwann cells with a recombinant retrovirus. Labelled Schwann cells arrived at the site of lesion two-three days after grafting, began to proliferate and expressed the peripheral myelin protein P0.[21-23,25] The migrating Schwann cells preferred pathways along the subarachnoid space,[24] ependyma, meninges and blood vessels but not the white matter.[22,26] Similarly to previous observations, migration in normal white matter never exceeded two-three mm while along other surfaces Schwann cells migrated at a speed of four mm/day as far as eight mm towards the lesion. In the less compact white matter of the shiverer mouse migration was more extensive but in uninjured animals little or no migration was found.[22] This finding together with the progressive directional migration of Schwann cells towards and within the lesion suggested that the lesion site triggers the migration of Schwann cells.[21,24]

Recent studies have shown that unmyelinating Schwann cells introduced into the CNS environment survive poorly and do not migrate unless they can myelinate axons.[26] Moreover, migrating Schwann cells do not seem to interact directly with myelin sheaths and oligodendrocytes.[27] These negative features of Schwann cells may be explained by multiple mechanisms acting between Schwann cells and the resident cell types of the CNS, such as the regulation of expression of polysialic acid-NCAM and N-cadherin.[26]

According to these results, grafted Schwann cells are a very potent cell population of the PNS capable of rapid and effective repair of demyelinated areas when transplanted into the CNS. They are able to migrate towards denuded axons and compete with less potent host oligodendrocytes for remyelination. On the other hand, remyelination by Schwann cells in the CNS is not entirely advantageous. Their aggressive penetration of the host tissue results in deteriorating changes of the microenvironment of the lesion site because Schwann cells may displace both host astrocytes and oligodendrocytes. This uncontrolled myelination may, in turn limit their use as potential cells to be grafted into demyelinated foci of CNS.

Transplantation of Olfactory Ensheathing Cells

Olfactory ensheathing cells (OECs) are resident in the olfactory nasal mucosa where they do not form myelin. They have common properties of both Schwann cells and astrocytes (for details see section on transplantation of OECs in chapter 4 and reviews by Ramon-Cueto and Valverde (1995)[28] and Bartolomei and Greer (2000)[29]). The rationale behind OEC transplantation into demyelinated areas of the spinal cord was that

A. although OECs normally do not produce myelin, they can myelinate axons in vitro[30] and

B. their astrocyte-like properties may enhance their myelinating capacity following transplantation.

It was assumed that the above features of OECs make them useful candidates for cell therapy to remyelinate demyelinated lesions. The most impressive series of grafting experiments were performed in the laboratories of Robin Franklin and Jeffery Kocsis.

Transplantation of OECs into an experimentally demyelinated rat spinal cord induced a pattern of remyelination very similar to that produced by co-transplantation of astrocytes and Schwann cells, i.e., a single OEC remyelinated a single axonal segment.[31,32] Again, the extent of remyelination was similar to that of the co-grafting experiments: OECs remyelinated axons along the whole lesion site without the need of additional cell types, and the grafted cells showed relatively good migrating capabilities.[33] The electrophysiological properties of the remyelinated axons revealed that their conduction velocity returned to near normal values and the conduction block was overcome.

In the above experiments it appeared relatively easy to produce an OEC line from a macrosmatic species, such as rat. It was uncertain, whether the same amount of tissue with the same properties as in the case of rodents, can be obtained from humans. Barnett et al (2000)[34] and Kato et al (2000)[35] have shown that human OECs obtained from olfactory bulbs and nerves can be purified and maintained in culture. The purified OECs were grafted into demyelinated areas of rat spinal cords where whey induced functional remyelination similar to that of rat OECs.

These studies have shown that human OECs may represent a new cell line for the development of transplantation strategy of CNS disorders and exhibit many properties of the rodent OECs including their capacity to form new myelin in the injured spinal cord.

Transplantation of Oligodendrocytes

Introduction

Oligodendrocytes are specific glial cells responsible for providing and maintaining CNS axons with insulating myelin sheaths, thus enabling rapid conduction of action potentials by axons. Accordingly, any harmful influence on the oligodendrocytes or any intrinsic defect may result in an incomplete or imperfect myelination or gradual loss of existing myelin sheaths. Recently it has become evident that axonal degeneration and/or spinal cord injury induces considerable oligodendrocyte apoptosis, thus reducing the number of glial cells that are able to remyelinate the lesion site.[36,37] There are well-known demyelinating diseases, such as the different forms of myelitis or multiple sclerosis. As spontaneous remyelination of demyelinated lesions in the mammalian CNS appears unsatisfactory, the studies on oligodendrocyte transplantation are of outstanding importance, not only because of the possible future use of oligodendrocytes to replace missing cells but also because oligodendrocytes together with astrocytes form a complex and well-balanced glial environment of axons and neurons. To restore such an environment by carefully selected proportion of glial cells naturally occurring in the CNS helps to understand the glia-neurone interaction in the CNS. Again, the spinal cord where long fibre tracts are situated on the external surface of the cord provides an excellent model for the study of migration and remyelination of grafted oligodendrocytes.

Even before successful transplantation of oligodendrocytes into the spinal cord was carried out, there was evidence to show that oligodendrocytes and their precursors are able to myelinate naked axons both in vivo and in vitro. In vivo experiments were carried out on hypomyelinated brains of mutants, such as the shiverer mouse. Hypomyelinated axons possess uncompacted myelin sheaths which lack the so-called major dense line due to the absence of a myelin protein of major importance (in the case of shiverer the Myelin Basic Protein, MBP).[38] Normal oligodendrocytes transplanted into a hypomyelinated brain of shiverer mice were able to overcome the defective shiverer oligodendrocytes and remyelinate naked axons.[39] Nevertheless, electron microscopic investigations revealed that transplanted and host oligodendrocytes competed for unmyelinated axons.[40] Similarly, if explants of ARA-C-treated CNS tissue was kept in vitro and was co-cultured with other tissue explants enriched in oligodendrocytes or cultured oligodendrocytes, remarkable remyelination occurred by the "grafted" oligodendrocytes and the explanted tissue became more mature.[41,42] Moreover, the explanted oligodendrocytes migrated towards naked axons.[42] The question arose whether oligodendrocytes taken from any part of the CNS are able remyelinate explants to the same extent. Interestingly, the myelinating capacity of dissociated oligodendrocytes or optic nerve explants superimposed upon ARA-C-treated cerebellar explants was less than that of organotypic explants or cultured oligodendrocytes.[41,43] The remyelination by "grafted" oligodendrocytes proved genotype-specific, i.e., shiverer oligodendrocytes formed shiverer-like myelin around normal host axons, while normal oligodendrocytes produced ultrastructurally normal myelin, suggesting that the myelin defect was due to a mutation of oligodendrocytes rather than abnormal axons. This is encouraging, for it suggests that remyelination may be achieved by donor oligodendrocytes.

Based on these results, a number of studies were initiated to investigate the remyelination process in the spinal cord. There were two main lines of investigations, one which induced remyelination by implanting solid pieces of embryonic nervous tissue or cell suspensions and the other used well-defined proportions of mixed glial cell cultures.

In the former case embryonic spinal cord[44,45] or brain[1,46] was grafted into the spinal cord of shiverer mouse or md rat (myelin deficient rat, x-linked myelin mutant) mutants and the remyelination was detected by immunocytochemistry and electron microscopy. Normal oligodendrocytes transplanted into shiverer or mdx rat CNS can be identified immunocytochemically by the presence of myelin basic protein or proteolipid protein, respectively. Normal myelin sheaths which displayed regular periodicity were formed around axons of mutants as early as 11 days after grafting.[44] Remyelinating oligodendrocytes appeared to have a tremendous capacity to migrate. They were found at remote sites from the injection and occasionally migrated up to six mm rostrocaudally.[45] Acute demyelination seemed to enhance the capacity of oligodendrocytes to achieve repair: if the shiverer cord was demyelinated by lysolecithin and the transplant placed one-three segments rostrocaudally from the lesion site, remyelination by host cells occurred nine days later and donor cells migrated three-eight mm towards the lesion.[46] However, the remyelination pattern was uneven. Most often only "patches" of normal myelin were observed.

At the edges of these remyelinated areas grafted oligodendrocytes often contacted abnormal host cells. In these areas there were no signs of astrocytic or microglial invasion of the grafted cells.[40,44] Similarly, freshly remyelinated fibres were never seen to be attacked by macrophages even if transplanted oligodendrocytes remyelinated axons in close vicinity of macrophages.[40] Nevertheless, the exact nature of these remyelinating cells was not clarified in these studies. It appeared likely that demyelination and transplantation enhanced the remyelination capability of host oligodendrocytes but host cells were not able to migrate over long distances as transplanted oligodendrocytes did. As such migration is primarily confined to glial progenitor cells[47] but not mature oligodendrocytes, it was suggested that, due to the relatively low number of identifiable precursor cells in the grafted tissue, several cell types of oligodendrocyte lineage including adult progenitor cells[48] could cooperate in different conditions to repair a demyelinating lesion.[45,46]

Factors that Determine the Success of Remyelination by Grafted Cells

A different environment for remyelinating oligodendrocytes was established by Blakemore and his coworkers. Lesions caused by application of a gliotoxic substance, ethidium bromide into the dorsal columns of the spinal cord are spontaneously repaired. However, when in addition oligodendrocyte precursors were eliminated by neonatal x-ray irradiation the lesions produced by the application of ethidium bromide were virtually free of glia and did not show spontaneous repair. Then subsequent remyelination is achieved by only the grafted glia and the extent of repair can be manipulated by the composition of the injected cells. Early experiments revealed that a mixed glial population (60-70% astrocyte, 25-40% oligodendrocyte and 5% Schwann cell) can remyelinate the demyelinated cord.[49] Both Schwann cells and oligodendrocytes remyelinated the naked axons of the dorsal columns and this process spanned up to six mm rostrocaudally from the site of the graft. However, where remyelination by Schwann cells occurred (peripheral remyelination) there were no oligodendrocytes or astrocytes present while numerous astrocytes were found in areas where remyelination was achieved by oligodendrocytes (central remyelination). This suggested that astrocytes might facilitate remyelination by co-transplanted oligodendrocytes and help oligodendrocytes in their competition with Schwann cells to myelinate so that remyelination by Schwann cells can be restricted to a minimum in favour of central remyelinating processes by oligodendrocytes[49] (for details on Schwann cell myelination see section: Transplantation of Schwann cells). This beneficial effect of astrocytes was confirmed in a series of studies where the proportion of oligodendrocytes and astrocytes grafted into the demyelinated cord was varied. Grafting of purified cultures of oligodendrocytes depleted of type-1 astrocytes resulted in remyelination by local Schwann cells but not by the grafted oligodendrocytes. Remyelination by grafted oligodendrocytes occured only when either the local Schwann cells were eliminated by x-ray irradiation prior to the ethidium bromide lesion or the oligodendrocyte cultures contained type-1 astrocytes.[50] Thus it became evident that oligodendrocyte cultures alone are not able to remyelinate demyelinating lesions and compete with Schwann cells which invaded the lesion. Remyelination by grafted oligodendrocytes was possible only if appropriate numbers of astrocytes were transplanted along with them. But why this difference in the myelinating capacity between Schwann cells and oligodendrocytes and how do astrocytes influence remyelination?

The explanation for the different behaviour, motility and myelinating capacity of oligodendrocytes and Schwann cells came from developmental studies on glial cell precursors. Both mature Schwann cells and oligodendrocyte precursors are motile and mitotic and respond to axonal mitogens.[40,47,51-53] Mature, myelin-forming oligodendrocytes are non-motile and non-mitotic,[48] though they possess a considerable capacity to myelinate axons when transplanted into md rat spinal cords.[54] It was suggested that the differentiation of oligodendrocytes,[55] the axon-glia interaction and hence the myelination of the axon is regulated by astrocytes. Astrocytes when grafted along with oligodendrocytes are capable to limit Schwann cell invasion and remyelination probably by forming the glia limitans in the grafted cord.[50,56,57] Moreover, astrocytes not only restrict Schwann cell incursion and remyelination but form an environment favourable for myelinating oligodendrocytes[50] but in order to establish this normal glial environment oligodendrocyte precursors have to be present.[58] However, in the absence of such co-operation, when no oligodendrocytes are co-transplanted with astrocytes, the grafted astrocyte population itself cannot prevent Schwann cell myelination.[56,58] This effect is probably due to the fact that in the absence of oligodendrocytes, type-1 astrocytes, like transplanted meningeal cells clump together and form a suitable surface for migrating and myelinating Schwann cells.[58-61]

Electrophysiological experiments revealed that grafting of glial cells which remyelinated the cord of md rats resulted in increased conduction velocity and improved frequency-response properties of the remyelinated axons indicating a proper functional recovery.[62]

The number of grafted oligodendrocytes also appeared critical for the outcome of remyelination. When mixed populations of cultured astrocytes and oligodendrocytes from early postnatal (P4) brains were injected into experimentally demyelinating lesions, considerable remyelination by either Schwann cells or oligodendrocytes occurred, depending on the proportion of grafted oligodendrocytes.[56] Grafts containing low proportions of oligodendrocytes (3-5%) resulted in Schwann cell remyelination while higher proportions of oligodendrocytes (10-15%) induced central (oligodendrocyte) remyelination but, again, only in the presence of co-transplanted astrocytes. Therefore, the number of oligodendrocytes and in particular their precursors appeared to play a crucial role in determining the extent of central myelination.

The Use of Oligodendrocyte Progenitors and Precursors

It is important to determine what type(s) of glial cells should be used for transplantation in order to replace missing glial cells or enhance the remyelinating capacity of the host cord. The rate of division and regenerative capacity of oligodendrocyte lineage cells decreases with differentiation, therefore earlier oligodendrocyte lineage cell types seem more appropriate candidates for successful transplantation. However, the question remains what degree of differentiation is needed to induce significant remyelination?

Very early neural progenitors isolated from the subventricular zone did not mature into myelinating oligodendrocytes following grafting into an unmyelinated lesion where they formed clusters of undifferentiated cells.[63] When the progenitors were induced to become committed to the oligodendrocyte lineage in vitro, significant remyelination has been achieved following grafting. Other precursors expressing the polysialylated form of the neural cell adhesion molecule (PSA-NCAM) differentiated in vivo into oligodendrocytes and astrocytes as they did in vitro, and unexpectedly they produced Schwann cells, too.[64] Oligodendrocyte progenitor cells lines produced more reliable results-they survived, migrated and remyelinated naked axons in a demyelinated spinal cord, inducing remarkable functional recovery after transplantation-but survived poorly in an intact spinal cord.[65,66] It should be noted, that the grafted progenitor cells were able to reverse the functional deficit only if the axonal loss was minimal.[66]

Further experiments provided evidence that the better survival of grafted oligodendrocyte progenitors in the x-ray-depleted spinal cords was due to the presence of vacant "niches" created by the depleted endogenous oligodendrocyte progenitors. Oligodendrocyte "survival factors" had no effect on the survival of progenitor cells.[67] Moreover, delayed availability of naked axons for the grafted oligodendrocyte progenitors also decreased their remyelinating capacity.[68]

The nature of the lesion also seems to influence the differentiation of glial progenitors. In situations where mainly oligodendrocytes are missing (for example md rat), transplanted bipotential O-2A cells (which are able to develop into astrocytes as well as oligodendrocytes under certain conditions) differentiate into oligodendrocytes, while in other pathological situations O-2A cells are able to produce both oligodendrocytes and astrocytes thereby reconstituting the damaged microenvironment.[69]

The behaviour of manipulated progenitor cell lines has also been studied when transplanted into a non-repairing spinal cord demyelinating lesion.[70] If O2A bipotential progenitor cells were immortalised with the ts A58-SV40T construct and grafted into demyelinated spinal cord they established a partially remyelinating environment with the presence of astrocytes while progenitor cells continuously treated with growth factors produced a well-myelinated environment even when only few astrocytes were present. Both treatments allowed the generation of sufficient cells to achieve remyelination but none of them was similar to the normal glial environment. However, the main problem with these engineered glial population was their failure to differentiate fully and the continued capacity to divide within the graft, in particular when growth factor-treated cultures were used. It also became evident that not all members of the oligodendrocyte precursor lineage possess equally good migratory and myelinating capacities.[71,72] Oligodendrocyte cultures enriched in early-phase precursors (A2B5$^+$, O4$^-$ cells) injected into the brains of shiverer rats migrated far from the injection site and remyelinated large areas of the CNS, while transplantation of later precursor types (O4$^+$, GC$^-$ and GC$^+$ cells) resulted only in scattered patches of myelinated areas. Bipotential precursor cells infected with a temperature-sensitive SV40T oncogene were used to compare their in vitro and in vivo characteristics (in culture and following transplantation).[73,74] The populations of precursor cells contained O4$^+$ and GFAP$^+$ (less then 10%) cells and at 38°C had a limited capacity to divide. This temperature-sensitive block of proliferation could be overcome by application of growth factors.[74] Early passages of these cell lines yielded myelin-forming oligodendrocytes and astrocytes following transplantation into a demyelinated cord and the cells forming myelin were often present in the areas demarcated by the astrocyte-like cells. Later passages of the cell lines were not able to form myelin. These experiments showed that carefully manipulated and controlled cell lines are able to engage in complex glia-axon interaction when grafted into demyelinated cords. Nevertheless, the most difficult problem is to find predictable correlation between the in vitro and in vivo behaviour of these cell lines.[71-73]

The role for host oligodendrocytes in the self-repair of chemically induced demyelination (without x-irradiation) was confirmed by Blakemore and his coworkers (1991,1995).[60,75] Xenogeneic (mouse) oligodendrocytes transplanted along with isogeneic rat astrocytes produced extensive remyelination in a demyelinated cord. The remyelination must have been produced by host oligodendrocytes, because, without immunosuppression, the mouse oligodendrocytes were rejected. However, the inflammation due to the rejection of the xenogeneic graft did not prevent, but augmented the repair by host oligodendrocytes, as host cells were remyelinating axons actually undergoing macrophage stripping of myelin sheaths (Fig. 3).[75] Similarly, experimental autoimmune inflammation reportedly promoted the survival and migration of grafted oligodendrocyte progenitors.[76] Given the fact that restorative processes are usually not augmented by inflammation and accompanying immunological reactions it would be very interesting to reveal the influences which act upon adult oligodendrocytes to trigger remyelination, because to promote remyelination by host oligodendrocytes would be the logical method to repair demyelinating lesions in the CNS.

Taken together, studies using experimental models of demyelination or hypomyelinating mutants have shown that mixed cultures of glial cells or newborn CNS fragments are able to achieve remyelination in areas of demyelination.[77-79] Unfortunately, the exact nature of remyelinating oligodendrocytes is not clear as remyelination may be due both to differentiated oligodendrocytes and/or undifferentiated precursors. Nevertheless, an increasing weight of evidence suggests that remyelination is chiefly due to migrating oligodendrocyte precursors though the role of adult oligodendrocytes cannot be neglected.[54,78] Possible therapies to achieve complete remyelination in demyelinating lesions are of extreme importance. At present it is not certain whether or not the best cure to remyelinate such lesions will be oligodendrocyte transplantation. An alternative solution and way of future research would be to encourage remyelination by adult host oligodendrocytes and this opportunity appears almost as promising as grafting cultured oligodendrocytes. Nevertheless, the glial environment and the interaction between astrocytes and oligodendrocytes seem crucial for remyelinating oligodendrocytes[59,78] therefore this aspect of human neuropathology needs to be further studied.

Remyelination Induced by Grafted Stem Cells

The limited availability of myelinating cells that can be used for transplantation urged researchers to study the myelinating capacity of stem cells. Stem cells used in these studies derived either from early embryos (embryonic stem cells)[80] or from later stages[81] including adult human brain-derived stem cells.[82] The stem cells were clonally expanded, embryonic stem cells needed induction to differentiate by retinoic acid. Although the cells maintained a versatile phenotype prior to grafting in culture, interestingly all the investigated stem cells turned into myelinating cell types in a dysmyelinated or amyelinated mutant rodent spinal cord after grafting. They produced either a CNS-type or a PNS-type myelin sheath indicating, that some of the neural precursor cells retain the capability to differentiate into cells with morphological characteristics of Schwann cells.[82] In one case the remyelinated axons were tested electrophysiologically and showed near normal conduction velocities. Briefly, it can be concluded that stem cells or clonal neural precursor cells are able to produce remyelination in the demyelinated spinal cord, although several questions remain to be addressed. Most importantly the controlled in vitro differentiation of the well-characterised neural precursor cell types should be precisely determined in order to avoid any unexpected outcome of the prospective transplantation trials in humans.

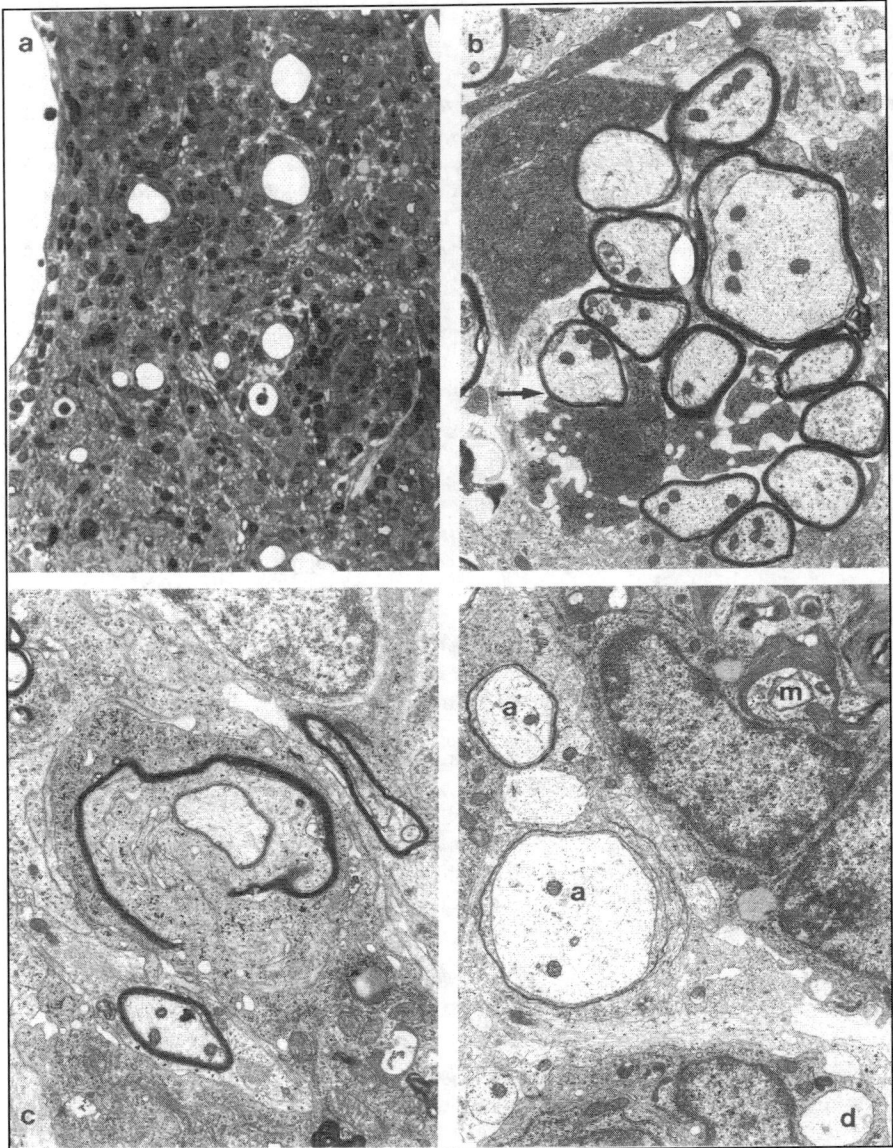

Figure 3. Following removal of immunosuppression lesions injected with mixed species cultures are infiltrated by inflammatory cells. When lesions from animals transplanted with mouse O2-A lineage cells and isogeneic astrocytes and maintained on cyclosporin for 2 weeks are examined 2 weeks after animals have been removed from immunosuppression the area of remyelination is heavily infiltrated with inflammatory cells. A) Large numbers of inflammatory cells are present within the lesion but few demyelinated axons are present. x450. B) Inflammatory cell processes (arrow) between remyelinated axons. x10,000. C) Macrophage processes engaged in the demyelinating processes known as "myelin sheath stripping". x13,000. D) A myelin debris filled macrophage (m) next to two axons (a) showing uncompacted thin myelin sheaths typical of early stages of remyelination. x11,000. Reproduced from GLIA (1995) 13:79-91, with kind permission from John Wiley and Sons Inc.

Transplantation of Astrocytes

Considerations that Led to Grafting Astrocytes into the Spinal Cord

Although it was suggested by Ramon y Cajal, that the failure of regeneration in the CNS of mammals was probably due to the glial scar formed after injury, only studies in the past few years have clarified some features of the effects of glia on axonal growth. Our knowledge about the regenerative capacity of the spinal cord of different species has been quite well-established in the 1960s, well before we knew much about the behaviour of reactive glial elements which play a role in the regulation of regenerative processes. According to classical views the regeneration of the adult mammalian spinal cord was non-existent, since after lesioning of the spinal cord neither function nor structure was restored. In contrast, in lower vertebrates, such as in lamprey, bony fish and goldfish return of the motor function of the transected spinal cord was observed.[83] Although functional recovery did occur, morphological restoration was incomplete. In the lamprey and young goldfish the regenerative axon growth observed was not prevented by the neuroglial scar[83] while in adult mammals the glial scar formed a barrier for regenerating axons.[84] The phylogenetic dichotomy on spinal cord regeneration was further complicated when significant regeneration was found in transected spinal cord of neonatal rats.[85] Also little gliosis was found in the injured spinal cord of neonatal rats (for more details see Chapter 3).[85]

These findings led to the conclusion that in the case of successful regeneration in primitive chordates and young goldfish the return of motor function (for example swimming) is only partially due to the reorganization of spinal cord circuitry and probably glial elements also contribute to the improvement of function either by facilitating or simply not preventing axonal growth or perhaps exerting a trophic action on spinal cord neurons. It has also been shown that mature astroglia inhibit whilst immature astrocytes facilitate axonal growth.[86-88] The reactive astrocytes are also not unanimously inhibitory in nature: reactive astrocytes secondary to a penetrating trauma (anisomorphic astrocytes) are more permissive for axonal growth than astrocytes responding to Wallerian degeneration (isomorphic astrocytes).[89] In contrast to the fact that reactive astrocytes are thought to inhibit axonal growth, they seem to facilitate neuronal survival in vitro through the release of soluble factors.[90] This effect could be of particular importance when reactive microglia which are known of their neurotoxic activity penetrate the areas of CNS injury.[90]

The putative trophic function of glial cells was also suggested when embryonic CNS tissue grafted into the brain has been shown to reverse deficits in spatial learning and memory[91] and improve behavioral assymetries caused by chemically induced catecholamine depletion.[92,93] Memory deficits could be improved just as effectively by grafted astrocytes from in vitro cultures as by fetal cortical transplants.[94] The interest in revealing the properties of grafted astrocytes resulted in a series of experiments where astrocytes were grafted alone or co-grafted with other glial cells into the spinal cord. The mammalian spinal cord proved a very good model for studying regenerative events as its elongated structure with several long fibre tracts allows the lesion as well as the subsequent intervention to be applied at various levels of the spinal cord.

Migration and Effects of Grafted Astrocytes

After successful attempts of grafting astrocytes into the brain, Bernstein and colleagues began to study the behaviour and migration pattern of the grafted and prelabelled astrocytes. In their experimental model E_{14} embryonic neocortex[95-99] or E_{14} spinal cord[100] was grafted into the thoracic spinal cord. The astroglial cells, prelabelled with Phaseolus vulgaris leucoagglutinin were identified by double-labelling for their GFAP content and were found to migrate as far as the lumbar spinal cord or dorsal column nuclei in the medulla.[95-96] The furthest distance from the graft where double-labelled astrocytes were found 90 days after transplantation was 55 mm suggesting a 0.72-0.76 mm/day migration rate. The primary direction of migration was rostrocaudal and migrating astrocytes mainly used the long fibre tracts, in particular the dorsal column tracts and glia limitans as pathways. Only limited migration was observed laterally in the spinal grey matter (up to 1 mm). Two functions were ascribed to the migrating astrocytes. Since grafted astrocytes appeared along the borders of the graft as early as seven days after transplantation when the grafted tissue itself contained no astrocytes, they were thought to have provided a migratory pathway for neurons of graft origin to penetrate the host tissue.[98] Indeed, when E[14] embryonic cortex was transplanted into the dorsal horn of the spinal cord, VIP (Vasoactive Intestinal Peptide) immunoreactive neurons, which are normally absent from the spinal cord, were observed close to motoneurons.[97] Interestingly, the reactive astrogliosis began one month after transplantation in the centre of the graft, suggesting that in this case the onset of gliosis corresponds to the termination of dendritic development. Such a delay in the onset of astrocytosis was thought to be sufficient for differentiation and migration of neurons.[97] The other function believed to be fulfilled by migrating astrocytes is merely trophic. Lesion to the fasciculus gracilis at cervical level results in loss of proprioception and deficits in hindlimb placement as well as atrophy of neurons in the dorsal column nuclei.[101,102] This deficit could be ameliorated by placing a solid piece of embryonic cervical spinal cord into the lesion cavity.[99,101,102] Nevertheless, no morphological reinnervation of the nucleus gracilis by grafted neurons occurred, i.e., the fasciculus gracilis was not repopulated by axons from the graft. Since astrocytes of graft origin were found in the nucleus, their possible trophic influence in maintaining the host neurons in the nucleus gracilis was suggested.[101,102] The nature of this trophic effect, however, remains to be determined. Attempts to do so revealed high levels of nerve growth factor (NGF) in the nucleus gracilis of injured animals 90 days after injury, but only slightly elevated NGF levels were found in animals in which the lesion was corrected with embryonic grafts.[103] Therefore NGF seems to be detrimental to neuronal maintenance and return of hindlimb functions, since increase of NGF was prevented possibly by migrated astrocytes and high NGF levels did not result in neuronal maintenance in the lesioned nucleus gracilis. The active presence of other factors exerted by grafted astrocytes is still far from proven.

Different results were obtained when cultured and purified astrocytes were grafted into the lesion cavity. Although they migrated to the nucleus gracilis, unlike astrocytes originating from solid tissue grafts they failed to prevent atrophy of the cluster neurons and improve hindlimb functions.[102] This finding was surprising because similar grafts of purified astrocytes placed into the cortex attenuated behavioral dysfunctions after frontal cortex lesion in rats.[94] Moreover, Wrathall et al[94] in 1985 reported that purified immature astrocytes grafted into the contused spinal cord have improved the functional deficits in rats. Wang et al (1995) grafted cultured astrocytes into the hemisected adult spinal cord.[105] Grafted astrocytes not only induced the decrease of the volume of scar tissue in the hemisected cord, but promoted the axonal growth around and into the scar tissue,too.[105] The grafted astrocytes migrated from the lesion site at a rate of 0.6 mm/day. It was not clear, whether these beneficial effects of grafted astrocytes resulted in improved functional recovery.

It cannot yet be determined which astrocytic factors may play role in maintaining or rescuing lesioned neurons: embryonic (immature) astrocytes may express cell surface molecules or produce trophic substances and both of these may be lost or modified during culturing or purification.[102] However, one should be aware that different purification and culture methods may result in loss of different surface molecules or may differentially alter the behaviour and maturation of astrocytes.

Another approach to transplantation of astrocytes into the spinal cord was used by Blakemore and his group.[57,59,106] These authors were interested in the repair of demyelinated lesions. The capacity of immature astrocytes to alter the microenvironment of an artificial demyelinating lesion and regulate the remyelination pattern achieved by Schwann cells and oligodendrocytes seems to provide a useful tool to achieve repair. Several in vitro and in vivo studies in the PNS[107] have demonstrated that astrocytes may form an environment suitable for growing axons and remyelinating oligodendrocytes. Accordingly, if type-1 astrocytes are transplanted into a lesioned spinal cord where the demyelinating lesion was produced by ethidium bromide and x-ray irradiation, the grafted astrocytes integrated with the damaged host tissue.[57] Moreover, a subsequent remyelination in the lesioned cord occurred which was due to the host oligodendrocytes. These cells normally have limited myelinating capacity, but the environment produced by grafted astrocytes may have induced them to increase their ability to produce myelin. Astrocytes may secret mitogenic (platelet derived growth factor) and possibly migratogenic factors which may stimulate adult host progenitor cells to remyelinate. The nature of this influence of astrocytes on the environment i.e., the extracellular matrix was studied by Franklin et al[106] In their experiments completely demyelinated lesions in the spinal cord were replenished by cultured and purified astrocytes. The grafted cell suspensions were carefully purified in order to remove oligodendrocytes and Schwann cells. The transplanted astrocytes were able to establish an astrocytic environment which contained an integrated astrocyte matrix, i.e., demyelinated axons surrounded with robust astrocytic processes. This indicated that the grafted immature astrocytes migrated, differentiated into a mature-like astrocyte type and integrated with the host tissue. Moreover, this environment formed by grafted astrocytes was very similar to the chronic demyelinated plaques of multiple sclerosis. In other studies the environment-forming effects of astrocytes derived from different sources were studied.[108] "Type-1" astrocytes derived from rat tissue cultures formed cords while "type-2"astrocytes, which differentiated from O-2A progenitor cells spread through a glia-free environment and thus showed greater capacity to fill glia-free lesions than tissue culture astrocytes. Similarly, progenitor-derived astrocytes filled infarcted areas of the spinal cord white matter more effectively than tissue culture astrocytes, although none of them was able to induce axonal growth into the reconstituted lesion site.

It appears from these results that astrocytes are responsible for the maintenance of the appropriate environment and physical framework within the CNS and facilitate the interaction between neurons and oligodendrocytes. However, the regulatory function of grafted astrocytes seems to be more complicated. Recent results indicate that astrocytes have a very strong stimulating influence on oligodendrocyte differentiation during development in vitro and are able to overcome the potent inhibitory effects of basic fibroblast growth factor.[55] Astrocytes also have a strong influence on the remyelinating capacity of co-transplanted oligodendrocytes and Schwann cells (for details see section: Transplantation of oligodendrocytes).

Transplantation of Macrophages into the Injured Spinal Cord

The CNS has been regarded as an "immunologically privileged" site, where there is little immunological survey of the brain and spinal cord. However, recently this view has been significantly modified, suggesting that the immune system, under specific circumstances, such as in case of infections or tumors is indeed involved into the response to these processes. The most important components of this limited immunological response are brain macrophages. Macrophages are known to remove debris after injury, to secrete cytokines and thus regulating mitogenic and chemotactic activities within the affected tissues (for review see Lotan and Schwartz 1994,[109] Schwartz et al 1999[110]). These potential "renewing" activities of the macrophages together with the limited inflammatory response of the CNS following injury suggested that grafting of macrophages into a lesioned spinal cord may be beneficial. Indeed, macrophages exposed ex vivo to regenerating peripheral nerves and then grafted into the lesion site of complete spinal cord transection induced some recovery in paraplegic rats and the functional recovery was accompanied by improved electrophysiological activity.[111] Morphologically descending fibres were seen passing the site of transection. Repeated transection of the fused

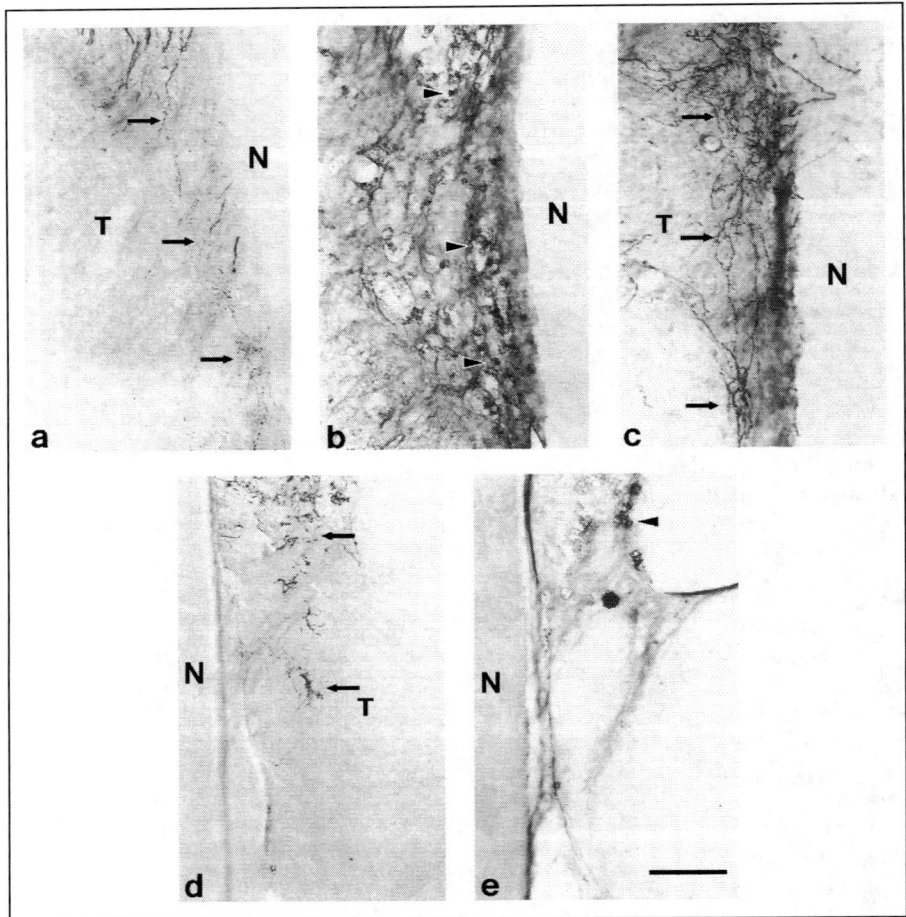

Figure 4. In these transverse sections the presence of CGRP-immunoreactive axons (arrows in A, C, and D) and ED-1 immunoreactive phagocytic cells (arrowheads in B and E) can be found in the transplant tissue (T) adjacent to the nitrocellulose implant (N). Adjacent sections (A and B) illustrate the extent of axonal growth and phagocytic cell activation following implantation of a TGF-β-treated implant, in contrast to d where there is sparse axonal growth from the dorsal roots apposed to an implant treated with MIF. In e phagocytic cells are found near the dorsal roots apposed to the fetal transplant; however, few ED-1 immunoreactive cells are aligned along the MIF-treated nitrocellulose implant. Bar, 100 μm for all figures. Reproduced from Experimental Neurology (1997) 148:433-443, with kind permission from Elsevier Science (Copyright 1997).

lesion site led to the loss of recovery, indicating that improved function was due to the re-established connections between proximal and distal spinal cord stumps.

Other studies have also shown the neurite-promoting activity of microglial grafts.[112-114] The above studies performed detailed immunohistochemical and molecular biological analysis and suggested that microglial cells may promote axonal regeneration

A.   by rapidly degrading myelin proteins and
B.   by promoting the synthesis of growth-promoting extracellular molecules.

This latter effect may be due to residing glial cells and Schwann cells that migrated into the lesion. Treatment of the grafted macrophages with Macrophage Inhibitory Factor prior to grafting resulted in reduced axonal growth-promoting activity of microglial grafts (Fig. 4).[112]

Although further thorough studies are needed to reveal the role of macrophages in spinal cord regeneration, the recent studies on macrophage/microglia transplantation suggest that grafted microglial cells are able to alter the microenvironment of the injured spinal cord and thus may promote axonal growth.

References

1. Gout O, Gansmuller A, Baumann N et al. Remyelination by transplanted oligodendrocytes of a demyelinated lesion in the spinal cord of the adult shiverer mouse. Neurosci Lett 1988; 87:195-199.
2. Jasmin L, Janni G, Moallem TM et al. Schwann cells are removed from the spinal cord after effecting recovery from paraplegia. J Neurosci 2000; 20:9215-9223.
3. Itoyama Y, Webster HdeF, Richardson EP et al. Schwann cell remyelination of demyelinated axons in spinal cord multiple sclerosis. Ann Neurol 1983; 14:339-345.
4. Ard MD, Bunge RP, Bunge MB. Comparison of the Schwann cell surface and Schwann cell extracellular matrix as promoters of neurite growth. J Neurocytol 1987; 16:539-555.
5. Blakemore WF. Remyelination of CNS axons by Schwann cells transplanted from the sciatic nerve. Nature 1977; 266:68-69.
6. Blakemore WF. The effect of sub-dural nerve tissue transplantation on the spinal cord of the rat. Neuropath Appl Neurobiol 1980; 6:433-447.
7. Blakemore WF. Remyelination of demyelinated spinal cord axons by Schwann cells. In: Kao CC, Bunge RP, Reier PJ, eds. Spinal Cord Reconstruction. New York: Raven Press, 1983:281-291.
8. Blakemore WF. Limited remyelination of CNS axons by Schwann cells transplanted into the sub-arachnoid space. J Neurol Sci 1984; 64:265-276.
9. Avellana-Adalid V, Bachelin C, Lachapelle F et al. In vitro and in vivo behaviour of NDF-expanded monkey Schwann cells. Eur J Neurosci 1998; 10:291-300.
10. Blakemore WF, Crang AJ, Patterson RC. Schwann cell remyelination of CNS axons following injection of cultures of CNS cells into areas of persistent demyelination. Neurosci Lett 1987; 77:20-24.
11. Kohama I, Lankford KL, Preiningerova J et al. Transplantation of cryopreserved adult human Schwann cells enhances axonal conduction in demyelinated spinal cord. J Neurosci 2001; 21:944-950.
12. Duncan ID, Aguayo AJ, Bunge RP et al. Transplantation of rat Schwann cells grown in tissue culture into the mouse spinal cord. J Neurol Sci 1981; 49:241-252.
13. Harrison BM. Remyelination by cells introduced into a stable demyelinating lesion in the central nervous system. J Neurol Sci 1980; 46:63-81.
14. Blakemore WF, Crang AJ. The use of cultured Schwann cells to remyelinate areas of persistent demyelination in the central nervous system. J Neurol Sci 1985; 70:207-223.
15. Iwashita Y and Blakemore WF. Areas of demyelination do not attract significant numbers of Schwann cells transplanted into normal white matter. Glia 2000; 31:232-240.
16. Brierley Cm, Crang AJ, Iwashita Y et al. Remyelination of demyelinated CNS axons by transplanted human Schwann cell: the deleterious effect of contaminating fibroblasts. Cell Transplant 2001; 10:305-315.
17. Duncan ID, Aguayo AJ, Bunge RP et al. Transplantation of cultured xenogenic Schwann cells into peripheral nerve and spinal cord of immunosuppressed mice. In: Kao CC, Bunge RP, Reier PJ, eds. Spinal Cord Reconstruction. New York: Raven Press, 1983:305-315.
18. Blight AR, Young W. Central axons in injured cat spinal cord recover electrophysiological function following remyelination by Schwann cells. J Neurol Sci 1989; 91:15-34.
19. Honmou O, Felts PA, Waxman SG et al. Restoration of normal conduction properties in demyelinated spinal cord axons in the adult rat by transplantation of exogenous Schwann cells. J Neurosci 1996; 16:3199-3208.
20. Blakemore WF, Crang AJ, Evans RJ et al. Rat Schwann cell remyelination of demyelinated cat CNS axons: evidence that injection of cell suspensions of CNS tissue results in Schwann cell remyelination. Neurosci Lett 1987; 77:15-19.

21. Baron-Van Evercooren A, Gansmüller A, Duhamel E. et al. Repair of a myelin lesion by Schwann cells transplanted in the adult mouse spinal cord. J Neuroimmunol 1992; 40:235-242.

22. Baron-Van Evercooren A, Duhamel-Clerin E, Boutry JM et al. Pathways of migration of transplanted Schwann cells in the demyelinated mouse spinal cord. J Neurosci Res 1993; 35:428-438.

23. Baron-Van Evercooren A, Gansmüller A, Clerin E et al. Hoechst 33342 a suitable fluorescent marker for Schwann cells after transplantation in the mouse spinal cord. Neurosci Lett 1991; 131:241-244.

24. Langford LA, Owens GC. Resolution of the pathway taken by implanted Schwann cells to a spinal cord lesion by prior infection with a retrovirus encoding ß-galactosidase. Acta Neuropathol 1990; 80:514-520.

25. Boutry J-M, Hauw J-J, Gansmüller A et al. Establishment and characterization of a mouse Schwann cell line which produces myelin in vivo. J Neurosci Res 1991; 32:15-26.

26. Iwashita Y, Fawcett JW, Crang AJ et al. Schwann cells transplanted into normal and x-irradiated adult white matter do not migrate extensively and show poor long-term survival. Exp Neurol 2000; 164:292-302.

27. Baron-Van Evercooren A, Avellana-Adalid V, Ben Younes-Chennoufi A et al. Cell-cell interactions during the migration of myelin-forming cells transplanted in the demyelinated spinal cord. Glia 1996; 16:147-164.

28. Ramon-Cueto A, Valverde F Olfactory bulb ensheathing glia: a unique cell type with axonal growth-promoting properties. Glia 1995; 14:163-173.

29. Bartolomei JC, Greer CA. Olfactory ensheathing cells: bridging the gap in spinal cord injury. Neurosurgery. 2000; 47:1057-1069.

30. Devon R, Doucette R. Olfactory unsheathing cells myelinated dorsal root ganglion neuritis. Brain Res 1992; 589:175-179.

31. Franklin RJM, Gilsson JM, Francheschini IA et al. Schwann cell-like myelination following transplantation of an olfactory bulb-ensheathing cell line into areas of demyelination in the adult CNS. Glia 1996; 17:217-224.

32. Imaizumi T, Lankford KL, Waxman SG et al. Transplanted olfactory ensheathing cells remyelinate and enhance axonal conduction in demyelinated dorsal columns of the rat spinal cord. J Neurosci 1998; 18:6167-6185.

33. Boruch AV, Conners JJ, Pipitone M et al. Neurotrophic and migratory properties of an olfactory ensheathing cell line. Glia 2001; 33:225-229.

34. Barnett SC, Alexander CL, Iwashita Y et al. Identification of a human olfactory ensheathing cell that can effect transplant-mediated remyelination of demyelinated CNS axons. Brain 2000; 123:1581-1588.

35. Kato T, Honmou O, Uede T et al. Transplantation of human olfactory ensheathing cells elicits remyelination of demyelinated rat spinal cord. Glia 2000; 30:209-218.

36. Warden P, Bamber NI, Li H et al. Delayed glial cell death following Wallerian degeneration in white matter tracts after spinal cord dorsal column cordotomy in adult rats. Exp Neurol 2001; 168:213-224.

37. Casha S, Yu WR, Fehlings MG. Oligodendroglial apoptosis occurs along degenerating axons and is associated with FAS and p75 expression following spinal cord injury in the rat. Neuroscience 2001; 103:203-218.

38. Privat A, Jacque C, Bourre JM, et al. Absence of the major dense line in the mutant mouse shiverer. Neurosci Lett 1979; 12:107-112.

39. Lachapelle F, Gumpel M, Baulac M et al. Transplantation of CNS fragments into the brain of shiverer mutant mice: Extensive myelination by implanted oligodendrocytes. II.Immunohistochemical studies. Dev Neurosci 1984; 6:325-334.

40. Gansmüller A, Lachapelle F, Baron-Van-Evercooren A et al. Transplantation of newborn CNS fragments into the brain of shiverer mutant mice: Extensive myelination by transplanted oligodendrocytes II. Electron microscopic study. Dev Neurosci 1986; 8:197-207.

41. Seil FJ. Tissue culture models of myelination after oligodendrocyte transplantation. J Neur Transplant 1989; 1:49-55.

42. Wolf MK, Brandenberg MC, Billings-Gagliardi S. Migration and myelination by adult glial cells:reconstructive analysis of tissue culture experiments. J Neurosci 1986; 6:3731-3738.

43. Stanhope GB, Wolf MK, Billings-Gagliardi S. Genotype-specific myelin formation around normal axons in cytosine arabinoside-treated organotypic cultures injected with normal or shiverer optic nerves. Dev Brain Res 1986; 24:109-116.

44. Rosenbluth J, Hasegawa M, Schiff R. Myelin formation in myelin-deficient rat spinal cord following transplantation of normal spinal cord. Neurosci Lett 1989; 97:35-40.

45. Duncan ID, Hammang JP, Jackson KF et al. Transplantation of oligodendrocytes and Schwann cells into the spinal cord of the myelin-deficient rat. J Neurocytol 1988; 17:351-360.

46. Gumpel M, Gout O, Lubetzki C et al. Myelination and remyelination in the central nervous system by transplanted oligodendrocytes using the shiverer model. Dev Neurosci 1989; 11:132-139.

47. Small RK, Riddle P, Noble M. Evidence for migration of oligodendrocytes-type 2 astrocyte progenitor cells into the developing rat optic nerve. Nature 1987; 328:155-157.

48. Wolswijk G. Noble M. Identification of an adult-specific glial progenitor cell. Development 1989; 105:387-400

49. Blakemore WF, Crang AJ. Extensive oligodendrocyte remyelination following injection of cultured central nervous system cells into demyelinating lesions in adult central nervous system. Dev Neurosci 1988; 10:1-11.

50. Blakemore WF, Crang AJ. The relationship between type-1 astrocytes, Schwann cells and oligodendrocytes following transplantation of glial cell cultures into demyelinating lesions in the adult rat spinal cord. J Neurocytol 1989; 18:519-528.

51. Crang AJ, Blakemore WF. Observations on the migratory behaviour of Schwann cells from adult peripheral nerve explants. J Neurocytol 1987; 16:423-431.

52. Kuhlengel KR, Bunge MB, Bunge R. Implantation of cultured sensory neurons and Schwann cells into lesioned neonatal rat spinal cord.I. Methods for preparing implants from dissociated cells. J Comp Neurol 1990; 293:63-73.

53. Kuhlengel KR, Bunge MB, Bunge R et al. Implantation of cultured sensory neurons and Schwann cells into lesioned neonatal rat spinal cord.II. Implant characteristics and examination of corticospinal tract growth. J Comp Neurol 1990; 293:74-91.

54. Duncan ID, Paino C, Archer DR et al. Functional capacities of transplanted cell-sorted adult oligodendrocytes. Dev Neurosci 1992; 14:114-122.

55. Mayer M, Bögler O, Noble M. The inhibition of oligodendrocytic differentiation of O-2A progenitors caused by fibroblast growth factor is overridden by astrocytes. GLIA 1993; 8:12-19.

56. Crang AJ. Blakemore WF. The effect of the number of oligodendrocytes transplanted into x-irradiated, glial-free lesions on the extent of oligodendrocyte remyelination. Neurosci Lett 1989; 103:269-274.

57. Franklin RJM, Crang AJ, Blakemore WF. Transplanted type-1 astrocytes facilitate repair of demyelinating lesions by host oligodendrocytes in adult rat spinal cord. J Neurocytol 1991; 20:420-430.

58. Franklin RJM, Crang AJ, Blakemore WF. Type 1 astrocytes fail to inhibit Schwann cell remyelination of CNS axons in the absence of cells of the O-2A lineage. Dev Neurosci 1992b; 14:85-92.

59. Blakemore WF. Transplanted cultured type-1 astrocytes can be used to reconstitute the glia limitans of the CNS:the structure which prevents Schwann cells from myelinating CNS axons. Neuropath Appl Neurobiol 1992; 18:460-466.

60. Crang AJ, Blakemore WF. Remyelination of demyelinated rat axons by transplanted mouse oligodendrocytes. GLIA 1991; 4:305-313.

61. Franklin RJM, Crang AJ, Blakemore WF. The behaviour of meningeal cells following glial cell transplantation into chemically-induced areas of demyelination in the CNS. Neuropath Appl Neurobiol 1992a; 18:189-200.

62. Utzschneider DA, Archer DR, Kocsis JD et al. Transplantation of glial cells enhances action potential conduction of amyelinated spinal cord axons in the myelin-deficient rat. PNAS 1994; 91:53-57.

63. Smith PM, Blakemore WF. Porcine neural progenitors require commitment to the oligodendrocyte lineage prior to transplantation in order to achieve significant remyelination of demyelinated lesions in the adult CNS. Eur J Neurosci 2000; 12:2414-2424.

64. Keirstead HS, Ben-Hur T, Rogister B et al. Polysialylated neural cell adhesion molecule-positive CNS precursors generate both oligodendrocytes and Schwann cells to remyelinate the CNS after transplantation. J Neurosci 1999; 19:7529-7536.

65. Franklin RJM, Bayley SA, Blakemore WF. Transplanted CG4 cells (an oligodendrocyte progenitor cell line) survive, migrate, and contribute to repair of areas of demyelination in x-irradiated and damaged spinal cord but not in normal spinal cord. Exp Neurol 1996; 137:263-276.

66. Jeffery ND, Crang AJ, O'Leary MT et al. Behavioural consequences of oligodendrocyte progenitor cell transplantation into experimental demyelinating lesions in the rat spinal cord. Eur J Neurosci 1999; 11:1508-1514.

67. Hinks GL, Chari DM, O'leary MT et al. Depletion of endogenous oligodendrocyte progenitors rather than increased availability of survival factors is a likely explanation for enhanced survival of transplanted oligodendrocyte progenitors in X-irradiated compared to normal CNS. Neuropathol Appl Neurobiol 2001; 27:59-67.

68. Blakemore WF, Chari DM, Gilson JM et al. Modelling large areas of demyelination in the rat reveals the potential and possible limitations of transplanted glial cells for remyelination in the CNS. Glia 2002; 38:155-168.

69. Franklin RJM, Blakemore WF. Glial-cell transplantation and plasticity in the O-2A lineage-implications for CNS repair. Trends Neurosci 1995; 18:151-156.

70. Crang AJ, Franklin RJM, Blakemore WF et al. The differentiation of glial cell progenitor population following transplantation into non-repairing central nervous system glial lesions in adult animals. J Neuroimmunol 1992; 40:243-254.

71. Warrington AE, Barbarese E, Pfeiffer SE. Stage specific, (O4⁺GalC⁻) isolated oligodendrocyte progenitors produce MBP⁺ myelin in vivo. Dev Neurosci 1992; 14:93-97.

72. Warrington AE, Barbarese E, Pfeiffer SE. Differential myelinogenic capacity of specific developmental stages of the oligodendrocyte lineage upon transplantation into hypomyelinating hosts. J Neurosci Res 1993; 34:1-13.

73. Trotter J, Crang AJ, Schachner M, Blakemore WF. Lines of glial precursor cells with a temperature-sensitive oncogene give rise to astrocytes and oligodendrocytes following transplantation into demyelinated lesions in the central nervous system. GLIA 1993; 9:25-40.

74. Barnett SC, Franklin RJM, Blakemore WF. In vitro and in vivo analysis of a rat bipotential O-2A progenitor cell line containing the temperature-sensitive mutant gene of the SV40 large T antigen. Eur J Neurosci 1993; 5:1247-1260.

75. Blakemore WF, Crang AJ, Franklin RJM. Glial cell transplants that are subsequently rejected can be used to influence regeneration of glial cell environments in the CNS. Glia 1995; 13:79-91.

76. Tourbah A, Linnington C, Bachelin C et al. Inflammation promotes survival and migration of the CG4 oligodendrocyte progenitors transplanted in the spinal cord of both inflammatory and demyelinated EAE rats. J Neurosci Res 1997; 50:853-861.

77. Duncan ID, Hammang JP, Jackson KF et al. Transplantation of Schwann cells and oligodendrocytes into the spinal cord of the myelin-deficient rat. J Neuropath Exp Neurol 1987; 46:351.

78. Blakemore WF, Franklin RJM. Transplantation of glial cells into the CNS. Trends in Neurosci 1991; 14:323-327.

79. Blakemore WF, Crang AJ, Franklin RJM. Transplantation of glial cell cultures into areas of demyelination in the adult CNS. Prog Brain Res 1990; 82:225-232.

80. Liu S, Qu Y, Stewart TJ et al. Embryonic stem cells differentiate into oligodendrocytes and myelinate in culture and after spinal cord transplantation. PNAS 2000; 97:6126-6131.

81. Hammang JP, Archer DR, Duncan ID. Myelination following transplantation of EGF-responsive neural stem cells into a myelin-deficient environment. Exp Neurol 1997; 147:84-95.

82. Akiyama Y, Honmou O, Kato T. et al. Transplantation fo clonal precursor cells derived from adult human brain establishes functional peripheral myelin in the rat spinal cord. Exp Neurol 2001; 167:27-39.

83. Bernstein JJ. Successful spinal cord regeneration: known biological strategies. In: Reier PJ, Bunge RP, Seil FJ, eds. Current Issues in Neural Regeneration Research. New York: Alan R Liss Inc, 1988:331-341.

84. Reier PJ, Houle JD. The glial scar: Its bearing on axonal elongation and transplantation approaches to CNS repair. In: Waxman SG, ed. Advances in Neurology Vol.47: Functional Recovery in Neurological Disease. New York: Raven Press, 1988.

85. Bernstein DR, Bechard DE, Stelzner DJ. Neuritic growth maintained near the lesion site long after spinal cord transection in the newborn rat. Neurosci Lett 1981; 26:55-60.

86. Kalderon N. Differentiating astroglia in nervous tissue histogenesis/regeneration: Studies in a model system of regenerating peripheral nerve. J Neurosci Res 1988; 21:501-512.

87. Kalderon N, Ahonen K, Juhász A et al. Astroglia and plasminogen activator activity: Differential activity level in the immature, mature and "reactive" astrocytes. Curr Iss Neural Reg Res 1988; 271-280.

88. Sumi SM, Hager H. Electron microscopic study of the reaction of the newborn rat brain to injury. Acta Neuropath (Berl) 1968; 10:324-335.

89. Mansour H, Asher R, Dahl D et al. Permissive and non-permissive reactive astrocytes: Immunofluorescence study with antibodies to the glial hyaluronate-binding protein. J Neurosci Res 1990; 25:300-311.

90. Giulian D. Reactive glia as rivals in regulating neuronal survival. GLIA 1993; 7:102-110.

91. Gage FH, Björklund A. Cholinergic septal grafts into the hippocampal formation improve spatial learning and recovery in aged rats by an atropine-sensitive mechanism. J Neurosci 1986; 6:2837-2847.

92. Dunnet SB, Björklund A, Schmidt RH et al. Intracerebral grafting of neuronal cell suspensions. IV. Behavioural recovery in rats with unilateral 6-OHDA lesions following implantation of migral cell suspensions in different forebrain sites. Acta Physiol Scand (Suppl) 1983a; 522:29-37.

93. Dunnet SB, Björklund A, Schmidt RH et al. Intracerebral grafting of neuronal cell suspensions. V. Behavioural recovery in rats with bilateral 6-OHDA lesions following implantation of migral cell suspensions. Acta Physiol Scand (Suppl) 1983b; 522:39-47.

94. Kesslak JP, Nieto-Sampedro M, Globus J et al. Transplants of purified astrocytes promote behavioral recovery after frontal lobe ablation. Exp Neurol 1986; 92:377-390.

95. Goldberg WJ, Bernstein JJ. Transplant-derived astrocytes migrate into host lumbar and cervical spinal cord after implantation of E14 fetal cerebral cortex into adult spinal cord. J Neurosci Res 1987; 17:391-403.

96. Goldberg WJ, Bernstein JJ. Grafted fetal astrocytes migrate from host thoracic spinal cord to lumbar cord and medulla. In Gorio et al., eds. Neural Development and Regeneration. NATO ASI Series Vol.H22. Berlin: Springer-Verlag, 1988b:525-538.

97. Connor JR, Bernstein JJ. Expression of peptides and transmitters in neurons and expression of filament proteins in astrocytes in fetal cerebral cortical transplants to adult spinal cord. Prog Brain Res 1987a; 71:359-371.

98. Connor JR, Bernstein JJ. Astrocytes in rat fetal cerebral cortical homografts following implantation into adult rat spinal cord. Brain Res 1987b; 409:62-70.

99. Bernstein JJ, Goldberg WJ. Rapid migration of grafted cortical astrocytes from suspension grafts placed in host thoracic spinal cord. Brain Res 1989a; 491:205-211.

100. Goldberg WJ, Bernstein JJ. Migration of cultured fetal spinal cord astrocytes into adult host cervical cord and medulla following transplantation into thoracic spinal cord. J Neurosci Res 1988a; 19:34-42.

101. Bernstein JJ, Goldberg WJ. Graft derived reafferentation of host spinal cord is not necessary for amelioration of lesion-induced deficits: Possible role of migrating grafted astrocytes. Brain Res Bull 1989b; 22:139-146.

102. Bernstein JJ, Goldberg WJ. Grafted fetal astrocyte migration can prevent host neuronal atrophy: comparison of astrocytes from cultures and whole piece donors. Restor Neurol Neurosci 1991; 2:261-270.

103. Bernstein JJ, Willingham LA, Goldberg WJ. Migrated fetal astrocytes modulate nerve growth factor expression in host nucleus gracilis of the medulla after grafting in third cervical hindlimb dorsal columns of the spinal cord. J Neurosci Res 1993; 34:394-400.

104. Wrathall JR, Pettegrew R, Castro M et al. Implantation of immature astrocytes into the contused spinal cord: Chronic effects on functional deficit and histopathology. Soc Neurosci Abs 1985; 15:1369.

105. Wang JJ, Chuah MI, Yew DTW et al. Effects of astrocyte implantation into the hemisected adult rat spinal cord. Neuroscience 1995; 65:973-981.

106. Franklin RJM, Crang AJ, Blakemore WF. The reconstruction of an astrocytic environment in glia-deficient areas of white matter. J Neurocytol 1993; 22:382-396.

107. Hall S, Gregson N, Rickard S. Interaction of regrowing PNS axons with transplanted aggregates of cultured CNS glia in vivo. J Neurocytol 1991; 20:299-309.

108. Blakemore WF, Olby NJ, Franklin RJM. The use of transplanted glial cells to reconstruct glial environments in the CNS. Brain Pathol 1995; 5:443-450.

109. Lotan M, Schwartz M. Cross talk between the immune system and the nervous system in response to injury: Implications for regeneration. FASEB J 1994; 8:1026-1033.

110. Schwartz M, Lazarov-Spiegler O, Rapalino O et al. Potential repair of rat spinal cord injuries using stimulated homologous macrophages. Neurosurgery 1999; 44:1041-1046.

111. Rapalino O, Lazarov-Spiegler O, Agranov E et al. Implantation of stimulated homologous macrophages results in partial recovery of paraplegic rats. Nature Med 1998; 4:814-821.

112. Prewitt CMF, Niesman IR, Kane CJM et al. Activated macrophage/microglial cells can promote the regeneration of sensory axons into the injured spinal cord. Exp Neurol 1997; 148:433-443.

113. Franzen R, Schoenen J, Leprince P et al. Effects of macrophage transplantation in the injured adult rat spinal cord: a combined immunocytochemical and biochemical study. J Neurosci Res 1998; 51:316-327.

114. Rabchevsky AG, Streit WJ. Grafting of cultured microglial cells into the lesioned spinal cord of adult rats enhances neurite outgrowth. J Neurosci Res 1997; 47:34-48.

CONCLUSION

The book gives an account of the normal anatomy and physiology of the spinal cord. It then summarizes the consequences of damage to the spinal cord and draws attention to the fact that in adult individuals of higher vertebrates there is little evidence of spontaneous repair after damage to the cord, while in lower vertebrates and immature animals of all species there is some though often limited repair and some recovery of function. Several approaches have been used to encourage functional and structural repair of the spinal cord of adult higher vertebrates. Some of the work was inspired by the better recovery of the damaged cord in young individuals and used various procedures of grafting of fetal tissue to encourage the recovery processes. Although the succcess of the various grafting procedures in terms of functional repair was limited, some encouraging information emerged.

The lessons learned also concern a more fundamental approach to the problem, in particular that of tackling the repair of the spinal cord in a more targeted manner. Many clinical observations indicate that the functional impairment of patients after spinal cord injury and their prospect for recovery is to some extend dependent on the site of the injury to the spinal cord. Thus it is said that patients with damage to the ventral part of the cord are less likely to recover then those with damaged dorsal part of the cord (see Chapter 2). If this was the case, then the experimental grafting procedures should target those parts of the cord that are known to leave the patient with the most severe disability.

In the case of grafting well defined populations of glial cell or neurones, the requirements are clearer. In the former case it is usually remyelination of pathways that is required; in the latter replacement of irretrievably lost neurones by fetal grafts is sought. In the case of neural grafts two different aims can be pursued. In one case grafted neurones can be used to replace various transmitters or other factors that are missing in the diseased CNS of the adult and which may promote recovery; in the other case replacement of specific and highly differentiated populations of cells are required by appropriate grafts of embryonic tissue. The problems facing us when attempting these manoevers, and the difficulties that need to be overcome before they can successfully be applied to treatment are discussed in the appropriate chapters.

Although the method of transplantation of glial or neuronal tissue into the spinal cord has been considered mainly as a means to repair damaged or diseased parts of the CNS, the results provided by these studies are of great importance to our understanding of several basic problems in neurobiology.

During normal development the different components of the spinal cord mature and interact with each other in a precisely orchestrated manner. To give just one example: motoneurones develop simultaneously with the appearance of particular trophic factors, and their synchronous development has often been linked so that it has been suggested that these factors are essential for motoneurone development. Experiments where embryonic motoneurones are placed into an environment where these factors are no longer present can answer the question whether such factors are indeed important for motonurone development or have another function in the developing spinal cord. Other issues such as the importance of a specific population of neurones for the survival of other cells in the developing CNS can also be studied using grafting procedures. Neurones of the red nucleus die if the spinal cord of neonates is

Transplantation of Neural Tissue into the Spinal Cord, Second Edition,
edited by Antal Nógrádi. ©2006 Eurekah.com and Springer Science+Business Media.

severed and they are deprived of their target. Grafting the appropriate tissue, i.e., spinal cord, can lead to their rescue, but other neuronal populations are unable to do so. Populations of embryonic motoneurones can to some extent repopulate a motoneurone-depleted spinal cord, but grafted cortical embryonic tissue cannot do this. Thus there is a certain degree of specificity within the spinal cord.

The potential benefits of the use of transplantation techniques for solving important clinical and theoretical problems are discussed.

While it is unlikely that any grafting procedure will achieve sucessful point-to-point reconnection of the severed parts of the spinal cord, it may be useful to consider how best to utilise the function of the intact circuitry below the lesion. This might be achieved by providing and grafting the appropriate mixture of neurones from supraspinal centres that may release the transmitters known to regulate the excitability of the spinal cord circuitry.

Replacement of motoneurones by embryonic grafts into a motoneurone-depleted spinal cord with the aim to repair paralysis has to consider in addition to the survival and integration of the grafted tissue a number of other problems accompanying the disease, such as the viability of the peripheral nerves to act as conduits for the new axons and the ability of the denervated muscles to recover. If grafting procedures in the spinal cord are to be of clinical use, these considerations have to be taken into account.

Index